WORLD BANK WORKING PAPER NO. 22

Causes of Deforestation of the Brazilian Amazon

Sergio Margulis

THE WORLD BANK
Washington, D.C.

Copyright © 2004
The International Bank for Reconstruction and Development / The World Bank
1818 H Street, N.W.
Washington, D.C. 20433, U.S.A.
All rights reserved
Manufactured in the United States of America
First printing: December 2003

 printed on recycled paper

1 2 3 4 05 04 03

World Bank Working Papers are published to communicate the results of the Bank's work to the development community with the least possible delay. The typescript of this paper therefore has not been prepared in accordance with the procedures appropriate to journal printed texts, and the World Bank accepts no responsibility for errors. Some sources cited in this paper may be informal documents that are not readily available.

The findings, interpretations, and conclusions expressed in this paper are entirely those of the author(s) and do not necessarily reflect the views of the Board of Executive Directors of the World Bank or the governments they represent. The World Bank cannot guarantee the accuracy of the data included in this work. The boundaries, colors, denominations, and other information shown on any map in this work do not imply on the part of the World Bank any judgment of the legal status of any territory or the endorsement or acceptance of such boundaries.

The material in this publication is copyrighted. The World Bank encourages dissemination of its work and normally will grant permission for use.

Permission to photocopy items for internal or personal use, for the internal or personal use of specific clients, or for educational classroom use, is granted by the World Bank, provided that the appropriate fee is paid. Please contact the Copyright Clearance Center before photocopying items.
Copyright Clearance Center, Inc.
222 Rosewood Drive
Danvers, MA 01923, U.S.A.
Tel: 978-750-8400 • Fax: 978-750-4470.

For permission to reprint individual articles or chapters, please fax your request with complete information to the Republication Department, Copyright Clearance Center, fax 978-750-4470.

All other queries on rights and licenses should be addressed to the World Bank at the address above, or faxed to 202-522-2422.

ISBN: 0-8213-5691-7
eISBN: 0-8213-5692-5
ISSN: 1726-5878

Sergio Margulis is Lead Environmental Economist in the Environment unit of the Latin America and Caribbean department at the World Bank.

Library of Congress Cataloging-in-Publication Data

Margulis, Sergio.
 Causes of deforestation of the Brazilian Amazon/Sergio Margulis.
   p. cm. -- (World Bank working paper; no. 22)
 Includes bibliographical references.
 ISBN 0-8213-5691-7
  1. Deforestation--Brazil. 2. Deforestation--Amazon River Region. 3.
Ranching--Environmental aspects--Brazil. 4. Ranching--Environmental aspects--Amazon
River Region. I. Title. II. Series.

SD418.3.B6M27 2003
333.75'137'0981--dc22
                                                                    2003063305

TABLE OF CONTENTS

Foreword vii
Second Foreword ix
Abstract xi
Acknowledgments xiii
Acronyms and Abbreviations xv
Executive Summary xvii

1. Motivation for the Study 1
2. **Deforestation and Land Use in Amazonia: Evidence of Large Scale Cattle Ranching** 5
   Temporal Trend of Deforestation 5
   Spatial Patterns at the Municipal Level 6
   Evolution of Land Use in Legal Amazonia 9
   Contribution of Large and Small Deforested Areas to Overall Deforestation 10
   Evolution of the Cattle Herd 13
   Socioeconomic Benefits from Deforestation: Early Evidence 14
3. **Different Frontiers and Their Economic and Social Dynamics: Determinants of Amazon Occupation** 17
   Expansion of the "Speculative Frontier" and the Deforestation Process ... 20
   Determinants of Land Occupation in Amazonia: An Econometric Model 25
4. **The Microeconomics of Beef Cattle Ranching in Amazonia** 29
   The Beef Cattle Economy in Amazonia: Brief Background 30
   Field Research 32
   Technical Parameters Adopted 33
   Production Costs: Pastures 36
   Net Income per Hectare 37
   Analysis of Yield 37
   Mathematical Modeling 39
   Some Simulations 41
   Final Considerations and Trends 43
5. **Social Costs and Benefits of Deforestation** 47
   Government Incentives: Subsidies as a Basis for Cattle Ranching and Deforestation? 48
   Estimate of Economic (Social) Costs of Deforestation in Amazonia 51
   Sustainable Alternatives: Comparing Costs and Benefits 55
6. **Conclusions and Recommendations** 59
   Main Results: Conclusions 59
   Recommendations 62

Annex: Socioeconomic Development in Brazilian Amazonia 67
References 73

TABLES

Table 1:	Evolution of Land Use in Legal Amazonia in Census Years (percent)	9
Table 2:	Evolution of Land Use in the cerrado	10
Table 3:	Property Rights in Legal Amazonia, in Census Years (percent of Geographic Area)	11
Table 4:	Distribution of the Number and Area Occupied by Establishments in Legal Amazonia, According to Size, Average for 1970-95	11
Table 5:	Average Size of Deforested Plots, 1996-99 (in percentage terms)	12
Table 6:	Percentage Contribution of Plot Sizes to Deforestation per State, Average 1997-99	13
Table 7:	Evolution of the Cattle Herd (1990-2000) (thousand animals)	13
Table 8:	Evolution of Per Capita Rural Income in the States of Legal Amazonia, 1970-95 (1995 US dollars)	15
Table 9:	Evolution of Beef Exports and Per Capita Consumption in Brazil (1995-2000)	15
Table 10:	Size of the Properties and Land Use Adopted in the Panels	34
Table 11:	Main Animal Production Indices as Indicated by the Panels	35
Table 12:	Comparison of the Parameters of the Panels with IBGE Data, Municipality of Paragominas (Pará)	35
Table 13:	Net Income per Hectare from Cattle (in Reals per year)	37
Table 14:	Internal Rates of Return (IRR) from Rearing, New-breeding and Fattening Systems	38
Table 15:	Internal Rates of Return (IRR) from New-breeding and Fattening System	38
Table 16:	Net Returns (in Reals per Hectare)	40
Table 17:	Absolute Deviations of Crops over the Past 10 Seasons (in January 2002 Reals per Hectare)	40
Table 18:	Results of Deforestation Tax Simulations	42
Table 19:	FINAM Fiscal Incentives by Sector (in 2000 US$ million)	45
Table 20:	Maximum Herd Numbers Benefiting from FINAM Incentives (a)	50
Table 21:	Rural Credit for Northern Region (in 2000 US$ million)	50
Table 22:	Summary of the Estimates of the Costs of Deforestation	54
Table 23:	Value of Renting the Land for Grazing in Legal Amazonia, 1998-2001 (US$ per ha/year)	55
Table A1:	Evolution of Rural and Urban Populations, GDP and GDP per capita in Legal Amazonia, 1970-95	71

FIGURES

Figure 1:	Accumulated Municipal Deforestation up to 2000	7
Figure 2:	Deforested Area Within Rural Establishments, 1995/96	8
Figure 3:	Cattle Densities (heads/km²) in 1975 and 1996	14
Figure 4:	Evolution of Transport Costs, 1980 and 1995	18
Figure 5:	Production Costs: Rearing—New-breeding—Fattening Panels	36

Figure 6:   Efficient Frontier: Risk and Income Combinations (Paragominas) 41

Figure A1:   Accumulated Distribution of Municipal Population in Legal Amazonia
             According to GDP per capita, 1970-1995 (in constant 1995 US$) 68

Figure A2:   Accumulated Distribution of Municipal Population in Legal Amazonia,
             According to Population with Insufficient Income, 1970-1991 69

Figure A3:   Average Life Expectancy in Municipalities of Legal Amazonia, 1970-91 69

Figure A4:   Infant Mortality in Municipalities of Legal Amazonia
             (deaths per 1000 live births) 70

Figure A5:   Cumulative Distribution of the Municipal Population of Legal Amazonia,
             According to Illiteracy Rates, 1970-91 71

Foreword

This study is part of a discussion process regarding the causes and dynamics of deforestation in the Brazilian Amazon. This initiative was launched by the Ministry of Environment in 2000, and the World Bank has been a partner in this process from the beginning. The objective of the debate has been to find coherence and a better understanding of the factors associated with the expansion of deforestation and the public policies attempting to arrest it.

Two underlying observations prompted this initiative: the first, that although the aggregate information on the deforestation rates of Legal Amazonia, published every two years by the National Institute of Spatial Research (INPE), contributed to the understanding of the evolution of the process, they did not allow for the formulation of short-term policies aiming at immediate interventions; the second, is that the interruption of fiscal and financial incentives prevailing in the 1970s, considered to be one of the determining causes of deforestation expansion, did not produce the expected results. Even in the absence of such incentives, deforestation was still growing.

These observations lead the Secretariat of Coordination of Legal Amazonia to invite specialists to debate and stimulate research, oriented towards changes in the policies implemented by the Ministry of Environment. It was necessary to disaggregate the data by States and municipalities in order to understand the dynamics of the different activities expanding in the region, and also to investigate the social and political forces acting in specific contexts, particularly those having greater influence on the growth of deforestation rates.

This work—"Causes of Deforestation of the Brazilian Amazon"—appears in this context. The central issue which it attempts to study is the role of cattle ranching—its dynamic and profitability—in the expansion of deforestation. If there are no fiscal incentives, other factors must make the activity feasible, justifying its continuous expansion in new forest areas. To answer this and other questions, the study analyzes the microeconomic behavior of cattle ranching, placing it in the broader context of the expansion of the logging and agricultural frontiers in one of the most important fronts of economic occupation of the region—Eastern Amazonia. In addition, the study compares the economic gains with the associated social and environmental costs, incorporating the socioeconomic dynamics of the social agents on this frontier into the analyses. It must be kept in mind that in the Amazon, agriculture and cattle ranching benefit from indirect gains coming from weak land titling, land grabbing, irregular labor contracts, and the continuous process of opening up of new forest areas. These are carried out at low cost by *posseiros* and small farmers, who prepare the land for more profitable enterprises which follow them.

Even though it started two years ago, the discussion presented in this study is extremely present. According to INPE's projection, between 2001 and 2002, deforestation in the Amazon went from 18,166 square kilometers to 25,476, the second largest increment since 1995. This fact in itself makes relevant the reading and discussion of this study, because of its contributions both to the diagnostic as well as the propositions of public policies. We hope this initiative stimulates new studies, more field work, and studies looking at feasible solutions, capable of influencing decisions not only in the government's environmental area but, more importantly, regional development policymakers.

Mary Helena Allegretti
Secretary of Amazon Coordination
Ministry of Environment

SECOND FOREWORD

Brazil's natural assets are legendary. The country is home to the largest rain forest biome in the world, the Amazon, containing by far the largest portion of remaining rain forest. The legal Amazon covers 60 percent of the Brazilian territory, with some 21 million inhabitants, or about 12 percent of the population, nearly 70 percent of whom live in cities and towns. Brazil also has the largest freshwater reservoir in the world, with the Amazon region alone containing up to one-fifth of the world's freshwater. Sustainable use of this enormous wealth would not only provide resources for the future, but also be a source of greater equity and poverty reduction since natural resources constitute a much higher proportion of the assets of the poor (some 80 percent) than of the rich.

The interest in sustainability is heightened by the Amazon deforestation rates. Of the original forest cover, 17 percent has been cleared, although at least a third of that is growing back. Its global value is seen in its rich biodiversity and the possible impact on climate due to its disappearance. Provisional data showing 25,400 sq. km. of deforestation in 2002, compared to an average of 17,340 sq. km. for the preceding ten years, illustrate the rising threat to key ecosystems. The near disappearance of Brazil's unique Atlantic forest earlier brings out the urgency of action. Some experiences worldwide and in Brazil with sustainable use of natural resources could serve as a basis for an environmental strategy with social inclusion.

The factors behind unsustainable resource use, however, are complex, and it is important to understand these sources if actions are to be effective. The Amazon region is challenged by a wide range of issues, including lack of consensus on development strategies; lack of adequate social services, infrastructure, and transportation; property right ambiguities and land use conflicts; rapid urbanization and poor quality of life in cities; inability to control deforestation and fires; role of indigenous people in development and environmental management; low institutional capacity and weak governance—and unmanaged expansion of cattle ranching and agriculture.

This study is a unique contribution in that it suggests that, in contrast to the 1970s and 1980s, when occupation of the Brazilian Amazon forest was largely induced by government policies and subsidies, much recent deforestation seems the result of medium- and large-scale cattle ranching, which is increasingly profitable and dominated by powerful agents. It emphasizes that it is essential to understand the strong private interests causing the increasing deforestation, and adopt policies that factor them into actions. It would also be essential to promote alternative and complementary business development that is more sustainable and equitable than at present. Controlling open access in the interior, while promoting the sustainable use of forested areas, seem key to avoiding as-yet poorly understood damage that may foreclose future options.

By bridging environmental and social policies, Brazil could in some form protect a sizable part of the existing Amazon forest, and improve the living standards and meet the aspirations of the local population. Overall, there is a recognition of the need for better policies and stronger institutions to manage natural resources. A growing domestic constituency supports sound environmental policies. Support for all this could be provided through public and private investments at the federal, state, and municipal governments, as well as policy analysis, knowledge exchanges, consensus building and the promotion of partnerships. A major opportunity for guiding World Bank contributions may be the federal *Amazônia Sustentável* program, with its linkages to the federal Multiyear Plan, and the Bank's new Country Assistance Strategy.

Vinod Thomas
Director, Brazil Country Management Unit
The World Bank

Abstract

The worldwide concern with deforestation of Brazilian Amazonia is motivated not only by the irreversible loss of this natural wealth, but also by the perception that it is a destructive process in which the social and economic gains are smaller than the environmental losses. This perception also underlies the diagnosis, formulation and evaluation of public policies proposed by government and non-governmental organizations working in the region, including the World Bank. The present work suggests that a fuller understanding is necessary with regard to the motivations and identity of the agents responsible for deforestation, the evaluation of the social and economic benefits from the process and the resulting implications of public policies for the region.

The objective of the report is to show that, in contrast to the 1970s and 1980s when occupation of Brazilian Amazonia was largely induced by government policies and subsidies, recent deforestation in significant parts of the region is basically caused by medium- and large-scale cattle ranching. Following a private rationale, the dynamics of the occupation process gradually became autonomous, as is suggested by the significant increase in deforestation in the 1990s despite the substantial reduction of subsidies and incentives by government. Among the causes of the transformation are technological and managerial changes and the adaptation of cattle ranching to the geo-ecological conditions of eastern Amazonia which allowed for productivity gains and cost reductions.

The fact that cattle ranching is viable from the private perspective does not mean that the activity is socially desirable or environmentally sustainable. Private gain needs to be contrasted with the environmental (social) costs associated with cattle ranching and deforestation.

From the social perspective, it is legitimate to argue that the private benefits from large-scale cattle ranching are largely exclusive, having contributed little to alleviate social and economic inequalities. The study notes, however, that decreases in the price of beef in national markets and increases in exports caused by the expansion of cattle ranching in Eastern Amazonia may imply social benefits that go beyond sectoral and regional boundaries.

From an environmental perspective, despite the uncertainties of valuation, the limited evidence available suggests that the costs of deforestation may be extremely high and possibly exceed private benefits from cattle ranching, particularly when the uncertainties of irreversible losses of genetic heritage (not yet fully understood) are incorporated. In this respect, activities such as sustainable forest management should be considered environmentally and socially superior. However, new policy instruments, funding mechanisms and monitoring and enforcement structures (that are difficult to implement) will be needed to make sustainable forest management a feasible alternative and to make ranchers internalize the environmental costs of their activities.

The key policy recommendations of the study are: (i) to acknowledge the private logic of the present occupation process of Brazilian Amazonia; (ii) to change the focus of policies towards cattle ranchers as the key driving force of deforestation, recognizing their interests and private economic gains; (iii) given the lack of knowledge about environmental costs and the uncertainties associated with the irreversiblity of present decisions, to formulate policies aimed at halting further expansion of the frontier in those areas which are still unaffected and encourage intensification of agriculture and cattle ranching in areas undergoing consolidation.

This study aims to stimulate and provide inputs to the debate on these themes, particularly between the government and the main agents of deforestation identified here (especially medium and large ranchers).

Acknowledgments

The attempt to better understand the economic rationale of deforestation of the Brazilian Amazonia is part of a process that started almost two decades ago, with the World Bank trying to help the government to elaborate alternatives for the sustainable development of the region. Few topics are so fascinating in the global environmental agenda. Probably because of this, few works of the Bank have received so many contributions and had such an in-depth review. Many colleagues were engaged in this process in order to ensure the clarity of the ideas and technical results presented here. In this sense, my personal thanks to Robert Schneider (in particular), Eustáquio Reis, Chris Diewald, Ken Chomitz, Joachim von Amsberg, John Redwood, and Luis Gabriel Azevedo.

The work was based on five "background papers" developed by teams headed by Geraldo Sant'Ana de Camargo Bastos and Sergio de Zen (ESALQ), Eustáquio Reis, Ronaldo Seroa da Motta (IPEA), Pablo Pacheco (IPAM/CIFOR), and Edna Castro (UFPA), whom I must congratulate and thank for the quality of the work.

Besides these, the list of people who I should also thank is too long, and I would never be able to do justice to all individually, or the institutions which they represent. From the Secretaria de Coordenação da Amazônia/MMA, with whom I discussed from the beginning the original motivation and the relevance of the theme: Mary Allegretti, Vanessa Fleischfresser, Mario Menezes, Katia Costa, and Brent Milikan.

World Bank colleagues under different conditions and circumstances: Adriana Moreira, Dan Biller, Claudia Sobrevila, Gregor Wolf, Ricardo Tarifa, Joe Leitmann, Daniel Gross, Robert Davis, Teresa Serra, Gunnar Eskeland, and Vinod Thomas.

From the field visits to Amazonia: Jonas Veiga, Judson Valentim, Alfredo Homma, Marisa Barbosa (EMPBRAPA), Marky Brito, Paulo Barreto, Eugenio Arima, Adalberto Veríssimo (IMAZON), Larissa Chermont, Ana Cristina Barros, Anne Alencar, Ricardo Melo, Cassio Pereira, Erivelto Lima, José Benatti (IPAM), Merle Faminow (IDRC), Jorge Orsi (SPRN-RO), Frederico Muller, Paulo Leite, Gina Valmorbida (FEMA-MT), and Philip Fearnside (INPA).

Coments, information and general technical data received from Diógenes Alves (INPE), Claudio Ferraz, Marcia Pimentel, Ajax Moreira (IPEA), Nilson Ferreira (IBAMA), David Kaimowitz, Sven Wunder (CIFOR), Hans Jansen (IFPRI), Dennis Mahar, Nigel Smith (U.of Florida), Jean Tourrand (CIRAD), Fernando Rezende, Gordon Hughes, Cornelis de Haan (consultants), and Alvaro Luchiezi (WWF).

Special thanks to FAO for funding part of the research, particularly to Henning Steinfeld, who also participated in technical discussions.

Thanks also to Lilian Santos, Mírian Felicio, and Karen Ravenelle for their editorial and printing support, and to John Penney for the translation.

I would also like to thank the various cattle ranchers, loggers, farmers and administrators who we interviewed in various cities in the field research, but who would be impossible to list individually—with them I have in fact learned a different perspective of the problem. Many thanks also for the cordial treatment given to us.

Finally, even though I have received help and constructive comments from so many colleagues, the results and opinions here presented are the author's sole responsibility, and do not necessarily reflect the points of view of the people above, of the institutions for which they work, and in particular of the World Bank.

Acronyms and Abbreviations

ADA	Amazon Development Agency
AML	Legal Amazonia
ARPA	Amazon Region Protected Areas Project
AU	Animal Unit
CEPEA	Center of Advanced Economic Studies
CGIAR	Consultative Group on International Agricultural Research
CSR	Remote Sensing Center of IBAMA
ESALQ	Agricultural University Luiz de Queiroz
FAO	United Nations Food and Agriculture Organization
FEMA	Environmental Agency of Mato Grosso
FINAM	Amazon Investment Fund
FNO	Constitutional Fund of the North
FPE	State Participation Fund
FPM	Municipal Participation Fund
FUNAI	National Indigenous Foundation
GDP	Gross Domestic Product
GEIPOT	Brazilian Transport Enterprise
ha	Hectare
IBAMA	Federal Environmental Agency
IBGE	National Geography and Satistics Institute
ICMS	Value Added Tax
IMAZON	Man and Environment Institute of Amazonia
INCRA	National Institute of Colonization and Agrarian Reform
INPE	National Space Research Institute
IPAM	Amazon Environmental Research Institute
IPEA	Applied Economic Research Institute
IRR	Internal Rate of Return
ITERPA	Land Institute of Pará
MCA	Minimum Comparable Areas
MMA	Ministry of Environment
MT	State of Mato Grosso
NPV	Net Present Value
PA	State of Pará
PPG7	Pilot Program for the Protection of Brazilian Tropical Rainforests
PROARCO	Fire Prevention and Control Program of Legal Amazon
PROBIO	Program for the Conservation and Sustainable Use of Brazilian Bioditerity
RO	State of Rondônia
SFM	Sustainable Forest Management
SIDRA	IBGE System of Automatic Data (Agricultural Research)
SP	State of São Paulo
SUDAM	Amazon Development Agency
UFPA	University of Pará
USP	University of São Paulo
UNCTAD	United Nations Conference on Trade and Development

EXECUTIVE SUMMARY

The World Bank and the Brazilian Government have for many decades discussed projects, strategies and the promotion of sustainable development policies for Amazonia. One of the fundamental questions raised in the course of this dialogue has been that of understanding the process of deforestation in the region. Despite these efforts, there still is no consensus about such basic questions as who the main agents of deforestation are and what their real incentives are. Recent studies by the World Bank and others have pointed to the low profitability of cattle ranching in the region—which leaves open the question how to explain the continued advance of the process even after fiscal incentives have been reduced or virtually eliminated.

Aside from this lack of knowledge, the process of deforestation in Brazilian Amazonia is widely regarded as detrimental to the environment, while producing limited economic and social gains. In addition to uncertainty about the precise extent of the environmental costs and losses caused by deforestation, perceptions of possible economic and social benefits lack a firm empirical and analytical basis. This has led to a range of opinions about the actual process of land occupation and deforestation. The following are some of the views commonly heard:

- The agents of deforestation operate with short-term planning horizons, and their activities are mainly based on forest nutrient mining;
- Cattle ranching is a low profit activity in Amazonia and only continues because it benefits from government credits and subsidies or because of prospective speculative gains;
- Small producers are important agents in the deforestation process;
- Timber extraction is one of the main causes of deforestation in the region;
- Roads are also causes of deforestation, and not consequences of the high agricultural and livestock potential of the region;
- Soybean cultivation is increasing rapidly in the *cerrado*, causing the agricultural frontier to expand into forest areas;
- Environmental costs measured locally, nationally and globally are so high that any activities which lead to deforestation are irrational;
- There are numerous alternative sustainable activities that could substitute cattle ranching and generate more substantial social, economic and environmental benefits.

The continuity and consistency of deforestation over the last few decades, after government incentives were substantially cut, suggest that, contrary to some of the views listed above, economic rationality does in effect underlie the process. Better understanding for this rationality is the main objective of the present study.

One of the basic premises of the study is that, although the economic potential of the region has not yet been thoroughly assessed, there is a need to seek options for sustainable development of the region. However, a strong sense exists that the economic activities that drive much of the current process of land occupation and deforestation—unsustainable logging and large-scale cattle ranching—fail to use the natural resource base in a way that maximizes net social benefits. Moreover, alternative models concerned with more sustainable and socially desirable uses of available resources *on the scale required* are not widely known. Therefore, the major dilemma faced by the Government is that while it is keen to exploit the region's vast potential in a sustainable way, it is not yet fully aware of the real extent of this potential. As a consequence, decisions that are taken now about the uses of the Amazon forest involve significant risks and uncertainty.

The main questions to be investigated include:

- Who are the principal driving agents and what is the rationale behind deforestation of Brazilian Amazonia?
- What is the true agricultural and cattle ranching potential of the Brazilian Amazonia and what are the natural limits to its expansion?
- What policies are appropriate in view of the uncertainties and risks involved in taking irreversible steps in ecological terms?
- Would it be possible or even desirable to "close" the frontier and focus on consolidating those areas which have already been occupied and expand cattle ranching in areas with greater potential?
- How to establish an alternative model "based upon the sustainable management of forest resources and its biodiversity" when this model has not yet been worked out on the necessary scale and while the "traditional" process of occupation continues at an increasing pace?
- Is the World Bank itself, which in light of its past experience in Amazonia decided to concentrate largely on protection and conservation of the forest, fulfilling its role as an agent of sustainable development?

The main contribution of the present study is to demonstrate that beef cattle ranching in Eastern Amazonia or on the consolidated frontier is highly profitable from the private perspective and that it produces rates of economic return higher than those obtained from the same activity in the country's traditional cattle ranching areas. In addition to the availability of cheap land, these higher returns are the result of surprisingly favorable production conditions—mainly rainfall, temperature, air humidity, and types of available pasture. The direct return on cattle ranching itself (excluding profits from the sale of timber) consistently exceeds ten percent. These are values potentially earned by the more professional and better capitalized ranchers operating on the consolidated frontier of Eastern Amazonia. Similar geo-ecological conditions in Western Amazonia, or in the areas where very dense forest cover predominates, make cattle ranching and agriculture practically (or totally) unviable there. This corroborates previous Bank studies which indicate that geo-ecological conditions are the main determinants of agriculture and cattle ranching in Amazonia.

This study also proposes that the financial viability of the medium and larger capitalized agents working in the consolidated frontier is the real motivating force behind the deforestation process in Brazilian Amazonia. The large number of intermediary agents who have lower opportunity costs and precede the larger cattle ranchers are probably directly responsible for much of the deforestation. Their activities are made viable in part by the assurance that they will be able to sell the cleared land at a future date, thereby covering their opportunity costs. Without this assurance of future land sales, the intermediary agents would have less incentive to deforest—at least on the present scale. The more professional and productive ranching ends the speculative cycle and the "nutrient mining" activities of the first agents and marks the beginning of the consolidation of the frontier.

Evidence on deforestation and land use in Amazonia presented in the study shows that cattle ranching is the main economic activity of the region and that the large and medium size operators are the major agents responsible for deforestation. The smaller agents are generally used to provide labor or to consolidate landholding by the so-called "warming" process (*esquentamento*), but these tend to make only a minimum direct contribution to deforestation. Regardless of the different motives, interests and economic strategies of the many social actors operating on the frontier, the end result of the land occupation process is most frequently the establishment of ranching activities. Cattle ranching enterprises now occupy nearly 75 percent of the deforested areas of Amazonia.

Agriculture does not compete with cattle ranching in the forest areas. Geo-ecological barriers are in general more restrictive in the case of agriculture—a prime example is the high precipitation levels in certain areas (above 2000 mm per year). Even in areas with less than this amount, rainfall is one of the major factors favoring cattle ranching making it predominate.

The study suggests that the high private profitability of ranching arising from the favorable geo-ecological conditions induces both deforestation and the building of roads. As long as the geo-ecological conditions remain favorable, there will be endogenous pressure to open more roads which will be privately built by cattle ranchers in the absence of government. If ranching were not profitable, the existence of roads *per se* or of a road network built with more geo-political aims in mind ("exogenous" roads) would not be the cause of the present level of deforestation or forest conversion. It is evident, however, that trunk roads constructed for geo-political purposes make ranching viable and therefore increase deforestation.

If cattle ranching is financially viable, the deforestation process does not only generate losses, as was assumed for the 1970s and 1980s. Even if private gains are less than the corresponding social and environmental costs, they do nevertheless generate income which is not only substantial but sustainable. In designing public policies for the region, it is vital to acknowledge this fact. *Cattle ranchers so far have played a minor role in the public debate and have not been taken into proper account in government policies. The study suggests policymaker's focus should shift from logging activities to ranching, since the latter plays a far more substantial role than the former in the process of deforestation.*

The financial viability of cattle ranching does not mean that public policies should support it. For such support, a social and environmental analysis of the costs and benefits of cattle ranching and deforestation should be carried out. There are three basic hypotheses regarding the balance of the social benefits and costs:

(A) Private benefits are less than social costs. In this case, deforestation is not defensible from a social standpoint. This is nowadays the generally accepted hypothesis regarding deforestation in Amazonia.
(B) The second hypothesis is that private and social gains exceed the social and environmental losses from deforestation. This hypothesis represents tradeoffs between economic activity and protection of the forest.
(C) As in (B) above, the third hypothesis is that social benefits are greater than the corresponding costs, but the net benefits are smaller than those obtained from other activities, such as sustainable forest management. In this case, public policies should not support cattle ranching.

In addition to the difficulties of measurement and monetary valuation, any such cost benefit analysis involves two further complicating factors. The first is that the externalities involved in the deforestation process in Brazilian Amazonia have a global dimension. Thus, social analyses must be made according to the perspectives of local communities, of the Brazilian population as a whole, and also of the international community (the latter being a direct or indirect beneficiary of the services of the Amazon forest). Results may vary according to the level of analysis. For example, would deforestation be acceptable if it were socially desirable for the local populations but not for the whole Brazilian population? Or would deforestation be defensible if it were in Brazil's best interests but not in that of the rest of the world?

The second factor is that Amazonia is an extremely large and diverse region, and results obtained from one subregion do not necessarily apply to others. As mentioned above, the rates of return indicated by the present study can be achieved by the more professional cattle ranchers in the "consolidated frontier" in Eastern Amazonia but not by the less advanced, average size operators throughout the whole Amazon region.

This study seeks to make a contribution in terms of a social evaluation of deforestation in Brazilian Amazonia by, on the one hand, identifying the main agents involved in the process, the economic motives behind their activities and their possible economic returns—a private evaluation of the process—and, on the other hand, by undertaking a monetary evaluation of the economic (social) costs of deforestation while making some comparisons with sustainable forest management—a social analysis. Although a full cost-benefit analysis has not been possible due to the lack of more substantive information, a number of different scenarios are presented and compared, allowing a number of policy recommendations for the region to be proposed.

From the social point of view, the potential benefits associated with deforestation can be seen first in terms of private profit obtained by the cattle ranchers and secondly in terms of the socio-economic improvements experienced by local populations or even by the Brazilian population as a whole. Since 1970, regional income has risen substantially. Rural income per capita in particular tripled on average—from US$410 in 1970 to US$1,417 in 1995. In the states with the highest deforestation rates, the increase was even higher: in Mato Grosso it rose from US$424 to US$4,311, in Pará from US$356 to US$1,436, and in Rondônia from US$712 to US$2,304.

Regional socio-economic indicators—such as literacy, infant mortality and life expectancy—show there has indeed been significant progress but it has been insufficient to reduce the gap in relation to the rest of the country. Moreover, the largest increase in regional income originated in the urban as opposed to rural sectors, which suggests that improvements in social conditions probably had little direct link with deforestation. In summary, even if the private gains from medium and large scale cattle ranching were significant, they appear to be largely exclusive, having contributed little to alleviate social and economic inequalities at the local level.

Even if local populations have not greatly benefited from the income generated by cattle production (and indirectly by deforestation), at the national level the price of beef has fallen continuously over the past few years, when 100 percent of the increase in cattle herd in Brazil resulted from the increased production in the main producer states of Amazonia—Pará, Mato Grosso, and Rondônia. Beef exports grew from 350,000 tons in 1999 to about 900,000 tons in 2002, representing approximately US$1 billion in export earnings.

As for the economic (social) costs of deforestation, despite the difficulty of quantifying and putting a monetary value on such costs, the study attempts to estimate costs so that comparisons can be made with the aforementioned benefits. The social costs of deforestation were estimated to be around US$100 per hectare per year. This figure is subject to great uncertainties because of the limitations of environmental valuation methodologies and data availability. A number of other effects of deforestation for which no precise data exists have also been omitted. However, the value exceeds the potential income to be derived from cattle ranching (about US$75 per ha/year), so that the latter could be compensated.

In spite of this balance, there are no transfer mechanisms to make effective such compensation, which would need to apply internationally since global externalities are involved. *This is an extremely important issue.* Current experiences in this respect are not encouraging, and there is an opportunity for the World Bank as well as the international donor community to play a role in this regard.

The study developed simulation models for agricultural activities, showing that producers are averse to risk and prefer to avoid specialization by adopting a combination of crops, pasture and forest where this is possible. The simulations show that producers would be prepared to accept relatively low sums (R$45 per ha/year—roughly US$15) as compensation for foregoing expansion of cultivated areas into the forest. These amounts could be as high as R$200 per ha/year in cases where risk aversion is regarded to be less pronounced (although in such cases producers would convert forest to agriculture and not to cattle ranching, under the model's assumptions). The simulations of imposing a tax on deforestation, instead of paying a compensation, suggest that producers would tend first to change the mix of crops, as opposed to reducing the amount of forest clearing. The difference between the two policies is who should bear the costs.

As for alternative activities, *sustainable forest management is shown to be less economic from the private point of view than cattle ranching.* It is also a very poorly-disseminated technique that is "sophisticated" in relation to unsustainable logging and cattle ranching, both widely practiced and perceived to involve very little risk. *However, from the social point of view the study shows that forest management is probably better on economic grounds.* It can be assumed that it is also better from the social and environmental points of view.

It should be noted that the analyses undertaken in the study must be viewed in context, especially as concerns the private economic viability of ranching which in principle reflects the specific conditions of the sub-regions studied. These conditions include mainly the level of professionalism of the ranchers, their scale of production and the amount of rainfall. Elsewhere in the region, results may differ from those presented in the study.

Recommendations

The study proposes the following policy recommendations for the region:

Information and Planning

- *It would be fundamental to recognize that cattle ranching in parts of Amazonia is a potentially profitable activity for producers and that this profitability is the basic driving force behind the deforestation process in these areas.* In this respect, the Government, the World Bank and the PPG7 should adjust the focus of their policies to change the incentives perceived by cattle ranchers. Acceptance of this thesis also implies acknowledging that important *tradeoffs* exist in the process of deforestation in Brazilian Amazonia.
- *Forest protection policies should, as a priority, be aimed at producers on the consolidated frontier* who are at the heart of the process and not at those on the speculative frontier. This does not mean that there should be no effort to enforce, fine and ban illegal operations by the agents on the speculative frontier.
- *The strategy should be to work with cattle ranchers and not against them.* While some of these agents may not be prepared or willing to negotiate, there are a number of more amenable major players interested in coming to some sort of compromise with the government and society in order to have their activities legalized.
- The authorities responsible for protection of the Amazon forest should also concentrate on loggers *but not in terms of deforestation* per se *but because disorderly, predatory and largely illegal timber extraction, as currently practiced, rules out the present or future possibility of sustainable forest management*, an activity which would be better than cattle ranching from an economic, social and environmental point of view.
- *Zoning should be encouraged as part of an educational and negotiating process between economic agents (including cattle ranchers) and the government, leading gradually to land occupation commitments in areas which are suitable from a social, economic and environmental point of view.*
- *Since a great deal of misinformation and uncertainty exists about the various factors associated with the deforestation process and expansion of the frontier, the risks involved suggest adopting conservative strategies.* With the heritage at risk in Amazonia, irreversible decisions involving potentially high social, economic and environmental costs should be avoided. In this sense, conservation initiatives should be encouraged and the World Bank will continue to support such initiatives through the ARPA, PROARCO, PROBIO, and a number of other projects sponsored by the Pilot Program.
- Among the most important factors analyzed in this study are the forests' environmental values and services and their associated possible social benefits. *It will be important to enhance knowledge about these values and services provided by the region. It is equally important to analyze more thoroughly the effects of transport costs on deforestation.*

Economic Instruments

- A system to introduce greater flexibility into the designation of strict conservation areas, such as *tradable development rights, could yield enormous ecological and economic benefits.* There is no reason why properties on fertile and productive areas should not be allowed to benefit from higher percentages of deforestation provided that they compensate this with

- additional areas of strict protection (legal reserves) in the ecologically richer areas. These areas would be identified by zoning initiatives (some have already been).
- *One of the classic economic solutions to the problem would be to tax deforestation* to oblige agents to internalize the environmental costs. The taxation simulations made in this study suggest that high taxes are necessary in order to significantly reduce deforestation. *An alternative would be to compensate agents for not deforesting.* The simulations made suggest that established ranchers willingness to accept for foregoing deforestation depends on their level of risk aversion. The choice of the instrument will depend on the decision on who should or could actually pay.
- In addition to the national interest, the international community also benefits from the environmental services of the forest. Even if the total sum of national and international benefits were potentially higher than the income from cattle ranching, *transfer mechanisms do not yet exist in practice and significant technical and political difficulties impede their implementation.* The World Bank has a potential role to play in assisting the Brazilian Government to examine relevant international initiatives and seek out appropriate partners.
- To date, the search for alternatives has been largely limited to sustainable forest management and small-scale pilot initiatives. Some of these are socially, environmentally and economically superior to cattle ranching. These efforts must be continued but *they do not yet compete with cattle ranching on the required scale.* Sustainable forest management which involves comparatively sophisticated techniques also has to compete with illegal logging and has proven difficult to be disseminated, calling for major government support.
- *The World Bank should review its largely conservation-oriented approach in relation to Amazonia over the past decade and focus more on the promotion of sustainable development* through productive activities with high social and economic benefits and low (or even zero) environmental impacts, represent alternatives for the future development of the region. Approval of a project in support of the National Forest Program, presently under discussion with the Federal Government, would be an excellent step in this direction.
- The regional fiscal incentives which benefited larger landowners in the past have been reduced and now tend to be better applied. *Social programs such as the preferential FNO and the INCRA settlement projects could bring more substantial ecological and social benefits, particularly for small producers.* This marriage of interests between environmental protection and support for traditional local populations is one of the top socio-environmental concerns of the new Brazilian government and should be supported.
- *Other economic instruments* which have been the subject of discussion for some time among Ministry of Environment, IPEA and World Bank officials might include: (i) the introduction of the Ecological Value-Added Tax (ICMS), (ii) the introduction of environmental criteria similar to those of the Ecological ICMS in the States and Municipalities Participation Fund (*Fundo de Participação dos Estados e Municípios*), (iii) reorientation of the criteria governing the award of fiscal or credit subsidies to promote sustainable activities and development of sustainable technologies and scientific research, (iv) strengthening the introduction of environmental criteria in the allocation of agricultural credit in the region, and (v) reviewing and fully eliminating existing subsidized credits for traditional cattle ranching in Amazonia.

Enforcing the Law

- *Whatever economic incentives are applied, there is a persisting need for greater surveillance and enforcement capacity. This will continue to be a major challenge* owing to the immense size of the region and the difficulties of working with the local stakeholders. Regardless of political determination, it will be difficult to reverse the inertial trend observed over several decades. However, recent experience in Mato Grosso demonstrates that this may be possible.

- *To ensure more effective action, a strategy of institutional cooperation is fundamental.* Agencies such as the MMA, IBAMA, ADA, the Ministry of Regional Development (*Integração Nacional*), the Ministry of Planning, INCRA, FUNAI, and the Federal Police—in addition to state governments—need to work together, agreeing common targets and defining individual functions.
- *Despite the political difficulties, the process of conceding property rights needs to be urgently and seriously reviewed and audited.* Here, without further analysis, it is difficult to identify the full network of interests involved. The results of not adequately addressing this issue, however, which are frequently associated with violence and fraud, are well known. These could be reversed if the agencies dealing with land occupation were to perform more effectively, bringing order to land use once and for all, protecting and supporting small producers and guaranteeing the integrity of public land and the natural and social assets of the region. The speculative gains surrounding land transactions in the region are very significant and the key government intervention is the legalization of property rights. The Federal Government working in partnership with the state governments should take decisive action on this issue.

CHAPTER 1

MOTIVATION FOR THE STUDY

Brazilian Amazonia has economic potential which is fundamentally based on the richness of its natural resources. The Government is faced with a significant dilemma: while it is anxious to capture this potential, it is unaware of its true extent. The new Minister of the Environment, in her inaugural speech, made a reference to this dilemma, suggesting that the region is viewed "…as a territory which almost naturally lends itself to the traditional expansion of the larger economy and less as the depository of an exceptionally wide biodiversity upon which its future development could actually be based." The principal questions include the following: What are the natural limits to the expansion of the agricultural frontier? Who are the agents and what is the logic of deforestation? Given the risks and uncertainty involved in taking irreversible steps in terms of environmental degradation and destruction of an area with a potential that is not yet entirely understood, what should be done? Would it be possible or desirable to "close" the frontier, consolidating the areas with some occupation and allowing the expansion of agriculture and cattle ranching in the areas with greater potential? How to implant an alternative model "based on biodiversity", when this model still raises doubts regarding its feasibility at the regional scale and while the "traditional" process of land occupation is progressing at an accelerating pace? Is the World Bank itself—which in light of its past experience in Amazonia put much of its efforts in recent years into forest protection and conservation—optimizing its potential role as a catalyst for sustainable development of the region?

In the search for answers to these questions, the causes and dynamics of deforestation in the Brazilian Amazon have been recurrent themes in a large number of studies over the past twenty years. These studies produced widely-accepted theses pointing to the key role played by cattle ranching as a cause of deforestation. At the same time, these studies also underscored the apparent economic irrationality and significant environmental costs of this process, affirming that deforestation was the result of activities which yielded low economic returns and depended largely on speculative gains or government subsidies. Dissatisfaction with the empirical bases of these conclusions, particularly in the present economic context of Amazonia, provided one of the motivations for the present study.

The study is also prompted by the recommendations of two of the most recent pieces of research done by the World Bank on the theme (Schneider et al. 2000 and Chomitz and Thomas 2000) as well as by those of an older study (World Bank 1991) which suggest that economic analysis of ranching in Amazonia is vital.

The present study was carried out in close collaboration with the Ministry of the Environment (MMA). It is in fact the product of a long term dialogue between the Bank and the government in search for conservation and sustainable development of the Amazon region. However, the present Report is the sole responsibility of the World Bank and its conclusions are not necessarily endorsed by the MMA.

The concept paper (concluded in September 2001) from which the present study emerged, reviewed critically the conventionally-accepted theses regarding the causes and dynamics of deforestation in Brazilian Amazonia. This gave rise to questions and alternative hypotheses requiring elucidation in the course of the study. The report contains a summary of the conclusions reached and outlines the evidence assembled in a number of previous research reports, and seeks answers to specific questions raised by the concept paper.

In this context, the study conducted a critical examination of a number of conventionally-accepted theses about the causes and dynamics of deforestation in Brazilian Amazonia. To a large extent they reflect a "common vision" about the deforestation process in Amazonia, for example:

- The agents of deforestation operate with short-term planning horizons and their activities are mainly based on forest "nutrient mining";
- Cattle ranching is a low-profit activity in Amazonia and continues only because it benefits from government credits and subsidies or because of prospective speculative gains;
- Small producers are important agents in the deforestation process;
- Timber extraction is one of the main causes of deforestation in the region;
- Roads are also causes of deforestation, and not consequences of the high agricultural and livestock potential of the region;
- Soybean cultivation is increasing rapidly in the *cerrado*, causing the agricultural frontier to expand into forest areas;
- Environmental costs measured locally, nationally and globally are so high that any activities which lead to deforestation are irrational;
- Numerous alternative activities could substitute cattle ranching and generate more substantial and sustainable social, economic and environmental benefits.

The totality of these theses would seem to imply that the deforestation process of Amazonia lacks economic rationale. It is seen to be a "lose-lose" process generating environmental destruction, limited economic benefits and trifling social gains. However, the fact that this process has continued over decades in most parts of the region, together with the difficulty of reversing it in the absence of a highly-coordinated and effective set of public policies, suggests that an underlying rationality to this process does in fact exist. An attempt to discover the nature of this rationality, as its title suggests, is the primary objective of the present study.

The concept paper suggested that the deforestation process in Amazonia does not merely consist of the devastation of new frontiers by activities that generate low economic and social returns but, on the contrary, the process is driven by an array of economic activities led by cattle ranching which generally produce substantial private economic gains.

In our understanding, the private financial analysis demonstrating high rates of return for cattle ranching under different conditions in Amazonia is perhaps the biggest contribution made by the present study. Of course, the financial viability of ranching does not mean that it should be supported by public policies. A social analysis of the process needs to be carried out, comparing the social costs and benefits of the activities associated with deforestation. It is important to know for example: (i) if the eventual income generated from activities in the wake of deforestation (especially ranching) is significant in terms of producing improvements in the living conditions of local populations (social welfare gains); (ii) whether the process is sustainable in the sense that it can be maintained over long periods without government subsidies; (iii) what the economic costs of deforestation are, and how they compare with the benefits.

Finally, even in the case that a social benefit-cost analysis showed a positive net marginal benefit from deforestation, justifying its adoption, it would still be necessary to compare it with alternative activities. These activities may prove to be "superior" to cattle ranching, in economic, social and environmental terms, justifying their adoption in place of cattle ranching. The key questions thus would be: do alternative activities which are ecologically, socially, and economically superior to cattle ranching exist? Would they substitute or complement cattle ranching? And the same set of questions regarding the economic justification of cattle ranching should be posed to them—their costs and benefits, social gains, and sustainability.

This study does not attempt to respond to all these questions and at all levels of analysis (private and social). Several topics, however, are studied in greater detail. Some of them are based on existing studies, while others employ the results of research financed as an offshoot of the present study. The main research framework was constructed around the comments and suggestions made during the concept paper review meeting. At that meeting, the following topics were identified as requiring further analysis:

- A comprehensive review of the microeconomics of cattle ranching in different parts of Amazonia, taking into account different production systems, different scales of production and focusing on the larger ranchers in particular;
- A detailed analysis to obtain a better understanding of the relationships between the pioneer occupiers and large cattle ranchers—Do contracts exist between them or do they indirectly build on one another's actions? Are the ranchers really behind everything?; and
- Statistics summing up deforestation in terms of plot size, location, background and trends, by regions or biomes, or according to some other appropriate regional and geographic measure.

A number of other pertinent questions were discussed at the review meeting but were not pursued in this study. However, reference is made to such topics as appropriate. These questions included alternative (sustainable) activities, law enforcement and issues concerned with legalization of property titles.

In order to verify and examine some of the hypotheses proposed in the concept paper and to attempt to respond to the above questions, five studies were carried out. The first—The Beef Cattle Economy and the Land Occupation Process in Amazonia by Geraldo Sant'Ana de Camargo Barros et al. (CEPEA/ESALQ-USP)—sought to produce a more detailed assessment of the (micro) economy of beef cattle ranching in Amazonia, focusing on some of the main production areas. The second—Deforestation in the Brazilian Amazon: a Review of Estimates at the Municipal Level by Pablo Pacheco (CGIAR/IPAM)—had as its main objective to evaluate and compare existing data on deforestation and analyze the data's consistency and reliability. The third—Actors and Social Relationships on the New Frontiers of Amazonia by Edna Castro et al. (UFPA)—basically aims to evaluate the link between agents on the more advanced deforestation fronts, in an attempt to understand the relationships between the larger and smaller agents. The fourth study—An Estimate of the Economic Cost of Deforestation in Amazonia by Ronaldo Seroa da Motta (IPEA/RJ)—was concerned with an evaluation of the costs of deforestation in Amazonia in order to compare them with its potential economic benefits. Finally, the fifth study—Land Occupation in Amazonia: Determinants and Trends by Eustáquio Reis and Ajax Moreira (IPEA/RJ)—specified and estimated econometric models which, at the municipal level, attempt to identify the relationship between the dynamic of the deforestation process and the expansion of the beef cattle ranching frontier and its social-economic implications. The five studies (all in Portuguese, except the second one) are available in the World Bank's Website (www.bancomundial.org.br).

In the course of the research new facts were learned, analyses were reassessed and several additional hypotheses emerged—some of which contradicted those originally proposed in the concept paper.

Objectives of the Study

The basic objective of this study is to better understand the dynamics and logic of deforestation in Amazonia. The main thesis, put forward in the concept paper, is that cattle ranching in significant parts of Amazonia, specifically in the more consolidated frontier regions of the Arc of Deforestation, is quite profitable from a private point of view, and thus comprises a key driving force behind this process.

The study also seeks to better understand the dynamic and links between cattle ranchers and other agents participating in the process—including loggers, small producers, local authorities, etc.—in order to understand their respective roles in the overall process.

Once this basic hypothesis regarding the substantial private gain to be obtained from cattle ranching is confirmed, the study analyzes other potential benefits from a social perspective as well as the economic costs involved in deforestation. This provides the basis for a social cost-benefit analysis of deforestation. While this is not fully achieved due to insufficient data, the numbers produced permit some insights into the size of the social costs involved and allow a number of tentative policy recommendations to be proposed.

In addition, though alternative sustainable activities that would be better than cattle ranching from a social perspective have not been analyzed, the study also compares the results obtained with published information on forest management.

Structure of the Study

The study is divided into five chapters following this brief introduction. Chapter 2 presents preliminary evidence demonstrating the importance of cattle ranching in the dynamic of the deforestation process. Given that ranching is on the whole predominant and that it continues to expand despite the elimination of Federal Government subsidies, the main conclusion is that it must therefore be economically viable from the point of view of private ranchers.

Chapter 3 attempts to distinguish between frontier agents in terms of economic size (value of assets or size of land) as well as their different economic motivations and strategies, whether speculative or productive. The chapter summarizes the results of one of the field studies which analyzed the social relationships between the agents on the "more advanced" frontier. In the second section, the econometric model seeks to identify the relationship between explanatory variables of the dynamic of the deforestation process and expansion of the ranching frontier.

Chapter 4 analyzes the microeconomics of beef cattle ranching in Legal Amazonia, summarizing the results of one of the research activities contracted under the present study. The findings of this research point to the private economic viability of beef cattle ranching in the areas studied, which present obvious comparative advantages.

Chapter 5 is an attempt to analyze from a social perspective the effects of the expansion of the regional economy in particular that based on cattle ranching. The chapter attempts to place a value on the economic (social) costs of deforestation. It also seeks to compare some of the costs and benefits of the process.

The final Chapter summarizes the main lessons learned and presents a set of recommendations for policies and actions that could be pursued by the Brazilian Government and the World Bank.

Chapter 2

Deforestation and Land Use in Amazonia: Evidence of Large Scale Cattle Ranching

Temporal Trend of Deforestation[1]

Analysis of spatial patterns and deforestation trends in Brazil's Amazon region suffers from the lack of a consistent, systematic empirical base. Brazil has however benefited significantly over the past decade from progress in the areas of remote sensing and satellite image processing. This has led to vastly improved knowledge of temporal and regional deforestation patterns in Amazonia. The key institutions responsible for primary data collection on deforestation through the employment of remote sensing techniques are INPE, IBAMA, and FEMA-MT. Each of these institutions applies the results of its monitoring effort to different purposes. Moreover, differences of interpretation of the images and differently defined parameters occasionally lead to substantial disparities between estimates.

Despite the comprehensive nature of the deforestation estimates, they are generally spatially aggregated. This makes it difficult to conduct an analysis of deforestation at the municipal or lower scale. More detailed assessments of the changes in vegetation cover over the past few years are temporally and regionally fragmented. Apart from remote sensing data, IBGE is the only source providing indirect estimates of deforestation on the basis of its land use surveys carried out every five years (with the exception of 1990) as part of the Agricultural Census.

Estimates even of the original forest cover show fairly wide discrepancies—356 million hectares according to FAO (1981), 379 million according to IBGE (1988), 409 million according to Skole and Tucker (1993), and 419 million hectares according to INPE figures (Faminow 1998). As for actual annual deforestation, while INPE estimated this to be 1.5 million hectares a year for the period 1978–1988, Skole and Tucker put this at an annual 2.1 million hectares.

Since 1988, INPE estimates have come to be regarded both at the central government and individual state level as the officially accepted statistics on deforestation in Brazilian Amazonia.

---

1. This section and the next are based on Pacheco (2002a), a background paper available at the World Bank's Web site (www.bancomundial.org.br).

INPE defines deforestation as "the conversion of areas of primary forest by human activities aiming at the development of agriculture/cattle ranching activities, as detected by orbiting satellites" (INPE 2000). On the basis of this definition, areas undergoing regrowth of secondary forests are excluded from the total gross annual deforestation figures, implying that an area once cleared will be considered so permanently.[2]

According to INPE, the total area deforested in Brazilian Amazonia increased from 15.2 million hectares in 1978 to 41.5 million hectares in 1990, 58.7 million hectares in 2000, and 60.3 million hectares in 2001 (the latter according to linear projections based on a sample of images).[3]

The spatial dynamics are well known. Since the beginning of the 1990s, studies have indicated that deforestation tends to be concentrated in a limited number of areas, with 76 percent of new deforestation occurring in only 49 Landsat images, and most of these in the so-called "arc of deforestation" (see Chapter 3). In 1998, 76 percent of all deforestation was concentrated in the states of Pará, Mato Grosso and Rondônia. Moreover, this percentage increased to 85 percent in 2000. Chomitz and Thomas (2000) suggested that 75 percent of all deforestation occurs within 25 km of municipal, state, or federal roads and 85 percent within 50 km of some of these roads.

Spatial Patterns at the Municipal Level

INPE Data

Two estimates made by INPE (Alves et al. 1997 and Alves 2000) suggest that deforestation is concentrated more intensively in the municipal districts to the south and southeast of Legal Amazonia (AML). They also suggest that deforestation is an inertial process in the sense that areas more likely to be deforested are next to those already cleared. Based on INPE data, Menezes (2001) estimated that in 1997,[4] 47 out of a total of 227 municipalities accounted for 50 percent of all deforestation in Mato Grosso, Rondônia, and Pará. Also, 139 municipalities covering an area of 123 million hectares accounted for 90 percent of the entire deforestation in the same three states, representing 77.4 percent of the total deforestation of Legal Amazonia.

IBAMA Data

IBAMA's main function is to carry out licensing and law enforcement operations. The IBAMA database carries information dating from 1996 referring to plots greater than one hectare. Forests at an advanced stage of secondary regeneration until 1996 are considered to be forested area. Eighty percent of the municipalities monitored by IBAMA fall within the "arc of deforestation"—the area boldly outlined in Figure 1 below—with the remainder dispersed throughout the states of Mato Grosso, Rondônia, and Acre. The main drawback of this data is the lack of satellite images for certain municipalities in certain years.

IBGE Estimates Based on Census Data

Estimates of changes in vegetation cover obtained from IBGE data are based on census surveys of agricultural establishments carried out at five year intervals since 1970 (except for 1990, when the economic censuses were not carried out and the demographic census was postponed until 1991).

---

2. INPE processes images on a 1:250.000 scale, which only allows for the identification of changes in cover greater than 6.25 hectares (corresponding to 1 mm^2 on the scale). Each image covers a 184 × 184 kilometer square, requiring 229 images to cover the whole of Legal Amazonia. Under cloudy conditions, INPE makes the assumption that the deforestation level is the same in the cloud-covered areas as that in the visible parts identified in the image (Faminow, 1998).

3. Approximately 44 satellite images, or around 20 percent of the total of 229 that cover the whole of Amazonia. These critical images are used to generate partial estimates of gross deforestation throughout the region. Deforestation interpolated for the period 2000-2001 is of the order of 1.7 million hectares (INPE, 2002).

4. Reference based on INPE figures, with images superimposed on IBGE map.

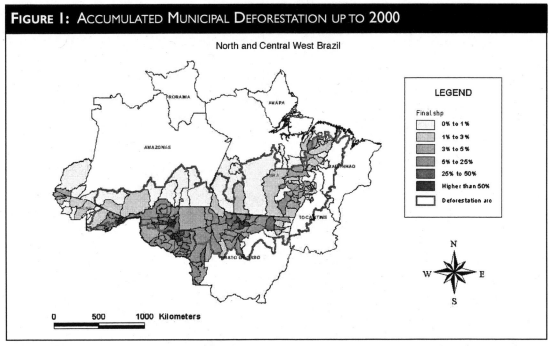

Source: Pacheco (2002a) based on IBAMA/CSR.

The 1995/96 Agricultural Census is the most recent census for which data are available.[5] The main advantage of the IBGE census information is that it provides estimates of the deforested areas in all the municipalities of Legal Amazonia over a long time span (Figure 2).[6] The disadvantage is the lack of precision of the deforestation estimates in cases where producers fail to make a direct declaration, resulting in estimates having to be obtained indirectly from statements regarding land use in the various agricultural/cattle ranching properties. In the following analyses, the areas are defined as the sum of the areas of annual and perennial crops, planted pasture,[7] planted forest, fallow areas and unutilized (but normally productive) land. The IBGE Census includes in principle all the agricultural/ranching properties in Legal Amazonia, but fails to give information relating to publicly owned or unclaimed land (*terra devoluta*). Deforestation estimates, therefore, have to rely on hypotheses about the extent of vegetation cover on publicly-owned or unclaimed land, as well as about preexisting vegetation cover on privately-owned properties.

---

5. In the Census carried out in 1996, IBGE – aiming to avoid providing information about planting and harvesting referring to different periods – changed the reference period from the calendar year (January–December) to the agricultural year (August–July). More importantly, data collection carried out between January and March (before the harvest) of the year following the census was assembled for the 1995/96 Census in the period between August and December 1996 (in the post-harvest period). This probably generated an underestimate of the number and area of agricultural/cattle ranching properties as the result of seasonal activities which are pursued by other groups of agents, particularly squatters, sharecroppers and tenant farmers (Eustáquio Reis, personal communication).

6. The map partially distorts the degree of deforestation occurring in the region, since it utilizes the area covered by agricultural/cattle ranching properties in the denominator in order to normalize the deforestation measurement. This is not entirely appropriate, since the municipal areas present significant variations in the size of geographical area appropriated or being used within the agricultural/cattle ranching properties.

7. Inclusion or not of natural pastures is a question to which no easy answer exists. On the one hand, it can be reckoned that they were originally savannah (*cerrado*), and therefore in the strict sense of the word do not fall into the deforestation category. On the other hand, it must be taken into account that the INPE estimates, for example, define forests by their physionomic features in the satellite images and therefore the felling that takes place in the savannah areas can be considered as deforestation. In this respect, the inclusion of natural pastures in the IBGE estimates makes for a better comparison with the INPE estimates.

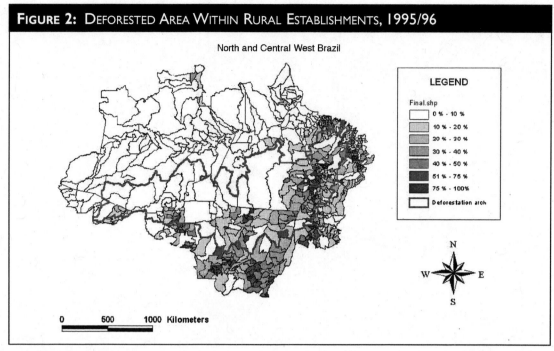

Source: Pacheco 2002a, based on 1995/96 IBGE Agricultural Census.

Summary

Reviewing the existing estimates based upon the main sources of deforestation data, INPE appears to overestimate net deforestation and possibly even gross deforestation, as suggested by IBAMA figures. The advantage of the INPE data is nevertheless their methodological consistency over a relatively long period of time.

The IBAMA data are based on a more detailed visual interpretation of images. This helps to ensure a much higher level of reliability. IBAMA's database however does not cover the whole of Legal Amazonia and its estimates could in fact be lower than actual deforestation in a number of municipalities because of the definition of forest used to measure accumulated deforestation in the baseline period, which treats the intermediate and advanced stages of secondary forest as if they were actual forest.

The indirect estimates based on IBGE census data provide an excellent source of information about deforestation for agriculture and cattle ranching purposes, despite the impossibility of distinguishing cleared areas from other types of vegetation. Analyses based on data from IBGE result in lower deforestation estimates than those indicated by remote sensing. This is due to the level of resolution, and to underestimates of the number of agricultural establishments. It should also be noted that some deforestation takes place outside the occupied areas and is not taken into account by the official censuses.

The data suggest that deforestation has not followed a linear trend. The phenomenon has in fact systematically increased since 1996. According to FEMA, there was an "encouraging" reduction in the level of deforestation in the state of Mato Grosso in the two-year period 2000-2001 compared with 1998-1999—approximately 32 percent—possibly due to the more aggressive monitoring, licensing and enforcement policies adopted by that state during the period (Fearnside, 2002). INPE data suggest that deforestation in the state of Mato Grosso in fact decreased by only 9 percent in 1999-2000, while in the two other key states (Pará and Rondônia) a marginal increase was recorded.

Fearnside (1993) suggests that much of the controversy surrounding deforestation estimates is caused by the way in which *cerrado* is differentiated from forests in the estimates. While INPE and IBAMA work exclusively with forest-type cover, FEMA includes all types of cover and does not differentiate cover types in its municipal data, only at the state level. Finally, IBGE Census surveys fail to differentiate between deforested land either in *cerrado* or forest areas (Andersen et al. 2002).

Evolution of Land Use in Legal Amazonia

The most basic statistics used to analyze the dynamics of deforestation in Amazonia is the evolution of land use in the region. These statistics are supplied by the Agricultural Censuses. Table 1 below shows that until 1970 the deforested areas used for agriculture and cattle ranching in Amazonia accounted for less than 3 percent of the total area of the region. Today, such areas account for over 10 percent. It is important to note that the denominator of the quotient is the *total area* of Legal Amazonia (5,075 million km^2), and not just the entire originally forested area, estimated at between 3,560 million km^2 (FAO 1981) and 4,190 million km^2 (INPE). The main change in land use is unquestionably the huge expansion of the area devoted to planted pasture, which by 1995 covered some 70 percent of the deforested areas. Assuming (a little exaggeratedly) that fallow areas are utilized basically for seasonal livestock rotation, pastures could account for the occupation of up to 88 percent of the deforested areas. Compared with 1970, 91 percent of the increment of the cleared area has been converted to cattle ranching.

More detailed analyses would be needed to elucidate changes in land use over time, the predominance of certain uses over others and the eventual stabilization of certain uses over time. These analyses have not been attempted here, but a number of econometric studies have been carried out on these lines (Reis and Margulis 1991; Pfaff 1997; Andersen and Reis 1997b; Ferraz 2000; Andersen et al. 2002; see also Kaimowitz and Angelsen 2000 for a major review). For our more limited purposes, available data strongly suggest that, in terms of the growth and spread of deforestation, cattle ranching is definitely the main economic activity associated with deforestation and that agriculture per se appears to have very little impact on deforestation.

This last observation raises the issue about the possible role of soybean as a cause of deforestation in the Amazon (Costa 2000, Becker 1999, Fearnside 2001). The bulk of converted land in the *cerrado* has been used for cattle ranching and not soybean production. The latter occupies a relatively small area of the anthropic *cerrado*, and the prospects for expanding into forest areas are limited (Costa, 2000; see also chapter 4). As can be seen in Table 2 below, the area of planted

TABLE 1: EVOLUTION OF LAND USE IN LEGAL AMAZONIA IN CENSUS YEARS (PERCENT)

	1970	1975	1980	1985	1995
Deforested areas	3.0	4.0	6.2	7.7	9.5
Total cropland	0.3	0.6	1.0	1.2	1.1
Planted pastures	0.7	1.4	2.6	3.8	6.6
Unused and fallow areas	2.0	2.0	2.6	2.7	1.8
Non-deforested areas	97.0	96.0	93.8	92.3	90.5
Public and protected areas	87.9	84.5	79.6	77.3	76.3
Natural pastures	4.0	4.5	5.1	4.7	3.6
Private forests (a)	5.1	7.0	9.1	10.3	10.6

(a) The areas covered by planted forests are 2 to 3 orders of magnitude smaller than the areas covered by natural forests.

Source: IPEA/DIMAC based on IBGE Agricultural Censuses.

TABLE 2: EVOLUTION OF LAND USE IN THE CERRADO (A)			
Area occupied by agricultural establishments, by type of use	1975 (1000 ha)	1996 (1000 ha)	Average annual growth (%)
Area occupied by agricultural establishments	110,798	124,314	0.5
Anthropic area (b)	34,695	64,487	3.0
Area occupied by crops	6,889	8,208	0.8
Area with planted pasture	16,053	49,207	5.3
Area of reforestation	586	757	1.2
Fallow area	356	671	7.4
Productive land not in use	10,815	4,643	–3.9
% of area in establishments	57.4	64.4	—
% of anthropic area in establishments	31.3	51.9	—
% of anthropic geographical area	18.0	33.4	—

(a) Based on 1975 data referring to homogeneous micro-regions and micro-regional data for the years 1995/96, with a view to securing approximate information about the core of the *cerrado*. Note it is not restricted to the Legal Amazonia.
(b) Anthropic area = the sum total of the area occupied by crops, planted pasture, reforested areas, fallow areas, and productive land not in use.
Source: Mueller (2002) based on IBGE Agricultural Censuses for 1975 and 1995/96.

pasture in the *cerrado* tripled between 1975 and 1995, while the crop areas increased by only 9 percent. Soybean, the most widely-grown crop in the *cerrado*, accounted for 6.3 million hectares in 2000, and represented only 10 percent of the converted area and 5 percent of the area in agricultural establishments. Soybean is increasingly grown in drier areas such as in parts of Mato Grosso or on abandoned pasture land. Regardless of the possible role of soybean cultivation, it can be inferred that cattle ranching must be sufficiently profitable and sustainable from a private perspective to explain much of the continuing growth suggested by these initial land occupation indicators.

With regard to the evolution of land use and land-holding in Amazonia, Table 3 below shows that there was substantial consolidation of legal property rights in those areas declared as cattle ranching establishments in the IBGE Agricultural Censuses (the percentage of establishments declared as "owned" by landowners increased from 70 percent to 95 percent), while the proportion of occupants—tenants, sharecroppers, and squatters—on the areas declined simultaneously (a reduction from 22 to 4 percent over the same period). As in the remainder of the country, a very pronounced land concentration was observed (Table 4). Both indicators are aggregated for the whole of Amazonia, thereby camouflaging regional variations. In general terms, however, there are no marked differences over time or inter-regionally.

Contribution of Large and Small Deforested Areas to Overall Deforestation

Satellite monitoring data provides information on the size of each deforested "plot," but it does not indicate whether each plot corresponds to a single establishment or landholding unit. In other words, it is possible to gauge from the satellite image the size of an entire deforested area, but it cannot be categorically affirmed that a specific area has been deforested by a single agent or

TABLE 3: PROPERTY RIGHTS IN LEGAL AMAZONIA, IN CENSUS YEARS (PERCENT OF GEOGRAPHIC AREA)

	1970	1975	1980	1985	1995
Owner	9.0	12.7	17.3	19.9	22.5
Tenant	0.8	0.5	0.8	0.5	0.2
Sharecropper	0.1	0.0	0.1	0.2	0.1
Squatter	2.8	3.0	3.4	2.1	0.9
Total	12.7	16.2	21.6	22.7	23.7

Source: IBGE, drawn by IPEA

TABLE 4: DISTRIBUTION OF THE NUMBER AND AREA OCCUPIED BY ESTABLISHMENTS IN LEGAL AMAZONIA, ACCORDING TO SIZE, AVERAGE FOR 1970-95

Size (hectares)	% of number	% of Area
< 2	31.9	0.3
2-10	23.6	0.9
10-100	29.9	9.8
100-1,000	13.0	27.9
1,000-10,000	1.5	31.4
> 10,000	0.1	29.6

Source: IBGE, drawn by IPEA

simultaneously by a number of different agents. INPE has been supplying statistics on the plots deforested since 1995. However, the resolution of the images does not permit one to discern cleared areas of less than 6.25 hectares. This rules out a precise assessment of the relative participation of the smaller agents in the process. For example, deforestation arising from "colonization" projects could take place in adjacent plots and therefore appear larger, thus leading to an overestimate of the relative contribution made by the larger agents. On the other hand, large- and medium-sized landowners could well clear isolated plots on the same property, overestimating participation by smaller producers.

A number of studies indicate the different roles played by the large and smaller agents. A certain degree of consensus exists regarding the significantly higher participation of the larger agents in the figures calculated for Amazonia as a whole (Cattaneo 2001, Faminow1998, Walker et al. 2000). Fearnside (1993) in particular suggests that 70 percent of the deforestation is caused by large cattle ranchers, while Homma et al. (1995) suggest that 50 percent of the deforestation is the result of subsistence farming. Chomitz and Thomas (2000) consider that establishments larger than 2000 hectares occupy over 50 percent of the converted areas. In any case, as indicated by Walker et al. (2000), significant subregional variations exist, and the causes and dynamics of deforestation differ significantly from one area to another, making it difficult to make generalizations. Nevertheless, as far as Legal Amazonia is concerned, it can be deduced that the larger and medium agents bear a significantly larger burden of responsibility for deforestation, as the data shown in Table 5 below suggest.

TABLE 5: AVERAGE SIZE OF DEFORESTED PLOTS, 1996-99 (IN PERCENTAGE TERMS)

Plot size (ha)	1997 Inpe (a)	1997 Ibama (b)	1998 Inpe	1998 Ibama	1999 Inpe	1999 Ibama
< 15	10.1(c)	7.5	10.9	10.5	14.8	9.8
15-50	23.1(d)	14.9	24.2	16.0	25.1	15.6
50-100	14.1	14.3	14.9	11.9	14.4	12.5
100-200	13.9	16.2	12.7	15.2	12.4	13.7
200-500	15.1	19.0	14.3	18.4	14.0	18.7
500-1000	9.4	14.3	9.5	12.7	8.4	12.4
> 1000	14.3	13.9	13.6	15.6	10.9	17.3
Total	100.0	100.0	100.0	100.0	100.0	100.0

(a) the whole of Legal Amazonia; (b) 170 municipalities for which data exists for all 3 years; (c) smaller than 10 hectares; (d) between 10 and 50 hectares.
Source: Adapted from Pacheco (2002a), based on INPE (2000) and IBAMA's site and IBGE.

The above table indicates that no large variations have occurred in the relative participation of each size category over the 1997-99 period, except, perhaps, for the decline of participation of the smallest agents (under 15 hectares). The discrepancy between INPE and IBAMA data regarding participation by each group is worthy of note. INPE data indicates that in 1999, approximately one sixth of the total deforestation took place in plots of under 15 hectares, whereas IBAMA figures tell a different story given that its data can identify cleared plots as small as one hectare. According to IBAMA, the contribution of plots of between one and three hectares accounted for about one percent of measured total deforestation in the 179 municipalities with the highest rates of deforestation in Legal Amazonia. In 1999, the smaller plots (less than 50 hectares) represented some 40 percent of total deforestation according to INPE, while IBAMA assessed this at only 25 percent. By contrast, INPE data indicate that the larger plots (over 200 hectares) were responsible for 33 percent of deforestation, while IBAMA placed this figure at 50 percent.

It is also interesting to discuss the IBGE results. IBGE estimates that the contribution to deforestation by establishments under 10 hectares amounted to 1.4 percent in 1995, while that in establishments under 50 hectares was 7 percent. Meanwhile, properties of over 1,000 hectares were responsible for 57.6 percent of the deforestation. These statistics vary significantly from the data produced by INPE and IBAMA and serve as a better indication of who the agents of deforestation are: the very large deforested plots may not make a larger contribution to deforestation estimates because they are simply the result of the sequential addition of smaller plots. Deforestation on the properties of over 1,000 hectares accounts for the highest proportion of that in Amazonia as a whole. The actual size of the deforested plots within these properties can be smaller.

Looking at the data on a more aggregated level, a variety of sub-regional patterns can be seen. The proportion of deforestation attributable to the smaller agents (measured by size of deforested plots) is larger in states such as Rondônia and Acre than in Mato Grosso, for example, where landholdings are much more concentrated (Table 6). The data suggest that even in Rondônia, known initially for smallholder land occupation, deforestation of areas under 15 hectares accounts for only 16 percent of total deforestation in that state. As confirmation of these trends, Fearnside (2002), using FEMA-MT data, suggests that only 2 percent of the areas deforested in Mato Grosso between 2000 and 2001 were smaller than 6.5 hectares.

TABLE 6: PERCENTAGE CONTRIBUTION OF PLOT SIZES TO DEFORESTATION PER STATE, AVERAGE 1997-99

Hectares	Rondônia	Pará	Mato Grosso	Other (a)	Total
< 15	16.1	10.5	5.1	15.3	9.5
15-50	25.4	8.9	10.9	18.7	14.0
50-100	17.6	13.1	11.0	13.7	13.1
100-200	15.7	15.0	14.9	16.1	15.2
200-500	14.3	19.9	21.1	16.1	19.0
500-1000	6.6	11.7	17.8	8.7	13.3
> 1000	4.3	20.9	19.2	11.4	15.9
Total	100.0	100.0	100.0	100.0	100.0
No. of municipalities	52	46	39	42	179

(a) Includes the states of Acre, Amazonas, Tocantins and Maranhão.
Source: Pacheco (2002a), based on INPE (2002) and IBAMA (www2.ibama.gov.br)

TABLE 7: EVOLUTION OF THE CATTLE HERD (1990-2000) (THOUSAND ANIMALS)

	1990	1995	1998	2000
Brazil	147,102	161,227	163,154	169,875
Mato Grosso	9,041	14,153	16,751	18,924
Pará	6,182	8,058	8,337	10,271
Rondônia	1,718	3,928	5,104	5,664

Source: IBGE

Evolution of the Cattle Herd

In addition to changes in land use, evidence regarding the expansion of the cattle herd in Amazonia corroborates the hypothesis that cattle ranching is a profitable activity in parts of Amazonia. As indicated in Table 7, the increase in the herd in Amazonia accounts for the bulk of the growth of the cattle herd in Brazil, suggesting that the cattle frontier has been pushed progressively towards the north. During the 1995-2000 period, for example, the equivalent of the entire increase in the national herd occurred in the three main producing states in Amazonia—Pará, Mato Grosso, and Rondônia. The remaining states tended to balance one another out in this respect, with some states presenting an increase in the herd and others a reduction. The average percent growth rates of the cattle population in these three states over the period 1995-2000 were 6.0 (Mato Grosso), 5.0 (Pará), and 7.6 (Rondônia), while at the national level the increase was only 1.1 percent.

The growth of the cattle herd in Amazonia was due partly to land freed up by deforestation and partly to more intensive ranching. Through regression modeling, the relationship between the expansion of the deforested area and the increase in the cattle herd (in animal units) was analyzed.

The regressions were made on the basis of variations in densities.[8] The results show that for the period 1970-95, the increase of one animal unit per hectare implied an average increase of 1.2 percentage points in the rate of deforestation in the area of a given municipality.[9] An interesting result is obtained by comparing the values between 1985-95 and 1970-85: the coefficient was reduced to less than half between the first and second periods (from 1.26 to 0.53), strongly suggesting intensified ranching. For each animal unit per hectare, the percentage of deforestation in the municipality associated with it declined from 1.24 to 0.53 percent. Intensified ranching can also be observed visually when the maps below, showing the geographical density of the cattle herd in Census Years 1970 and 1996, are compared (Figure 3).

Socioeconomic Benefits from Deforestation: Early Evidence

The data thus far presented demonstrate the dynamism of the rural economy in Amazonia in the last decades. It would be critical, however, to analyze whether the activities which lead to deforestation in the region generate more welfare gains than the social, environmental and economic costs involved and whether these gains benefit in some relevant way the poorer population groups.

The annex to this report summarizes statistics which were developed as part of this study with the specific objective of analyzing the evolution of socioeconomic indicators of Amazonia. The original objective was to assess the possible social gains associated with the process of land occupation and deforestation in the region. However, most of the available socioeconomic indicators do not distinguish between urban and rural groups, making it almost impossible to attribute

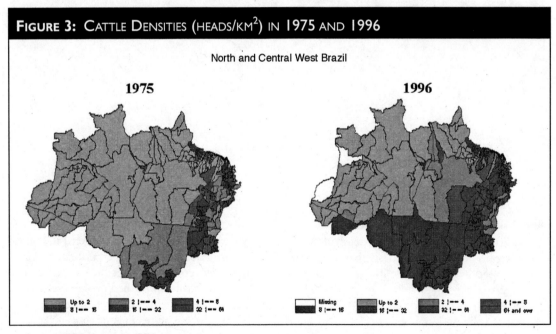

FIGURE 3: CATTLE DENSITIES (HEADS/KM^2) IN 1975 AND 1996

Source: IBGE, drawn by IPEA

---

8. With the aim of eliminating the effects of spurious correlations, which might result from the difference in the sizes of the municipal areas which would probably tend to slope the angle regression coefficients towards 1 (unit).

9. Note that it is not the percent variation in the deforestation rate, but rather the percent point variation of that rate (i.e., the rate goes from, say, 8 to 9 percent, as opposed to increasing by 1 percent).

TABLE 8: EVOLUTION OF PER CAPITA RURAL INCOME IN THE STATES OF LEGAL AMAZONIA,[A] 1970-95 (1995 US DOLLARS)

State	1970	1975	1980	1985	1995
Acre	547	492	760	894	1,089
Amapá	294	404	673	817	2,467
Amazonas	488	554	726	1,145	1,280
Maranhão	270	376	434	404	509
Mato Grosso	424	629	1,307	1,202	4,311
Pará	356	473	784	909	1,436
Rondônia	712	755	832	1,139	2,304
Roraima	785	1,126	1,121	1,102	1,202

(a) —except Tocantins
Source: IBGE, drawn by IPEA.

TABLE 9: EVOLUTION OF BEEF EXPORTS AND PER CAPITA CONSUMPTION IN BRAZIL (1995-2000)

	1995	1996	1997	1998	1999	2000
Exports (1000 t.eq. carcasse)	277.9	274.8	300.6	381.1	560.0	586.8
Annual per capita consumption (kg/yr)	35.6	38.4	36.0	34.4	34.7	35.7

Source: Cepea/Esalq/USP (2003).

improvements in socioeconomic indicators specifically to rural activities, and indirectly to deforestation. While the causality is unclear, the indicators presented in the annex demonstrate overall that there have been significant improvements in the socioeconomic conditions in the region, although these improvements have basically kept pace with the rest of the country, so that the gap in relation to it has not really narrowed.

Despite these caveats, some indicators are available for specific rural groups and can thus be more easily associated with the deforestation process and the activities which follow it. Two such indicators are presented here. The first has to do with rural income. Table 8 shows the significant rise of GDP per capita in the rural areas in almost all the states of Amazonia. With the exception of the state of Amapá (which started from a very low base), the "development leaders" were precisely the three states with the highest levels of deforestation (Pará, Mato Grosso, and Rondônia). Between 1985 and 1995, the very high growth registered in the state of Mato Grosso could be attributed mainly to the expansion of soybean in the *cerrado* region of the state. Figures for per capita GDP, of course, do not inform about effective appropriation of income by poorer or local population groups.

The second represents social benefits accruing to the national population. The expansion of the cattle herd in Amazonia has been largely responsible for beef prices dropping in real terms recently (as already indicated in Table 7, it represented nearly 100 percent of the growth of the national herd since 1995). It has also allowed beef exports to expand significantly. This study did not make a detailed evaluation of such benefits, but Table 9 illustrates the recent evolution of beef exports and the annual per capita beef consumption in the country.

In value terms, exports grew from less than US$500 million in 1995 to US$1 billion in 2000. Per capita consumption of beef did not increase. This does not necessarily mean that the entire increase in production has been exported, since the types of meat exported and consumed domestically are different. Analysts suggest further that Brazil has a major potential of expanding its share in the global beef markets. While beef exports represent important foreign exchange earnings, the question is whether these and other benefits justify the social and environmental costs of deforestation. These are analyzed in chapter 5.

CHAPTER 3

DIFFERENT FRONTIERS AND THEIR ECONOMIC AND SOCIAL DYNAMICS: DETERMINANTS OF AMAZON OCCUPATION

From an economic point of view, the Brazilian Amazonia can be characterized by abundant natural resources—arable land, mineral deposits, natural forests, etc.—which as yet are far from fully explored and much of which remain outside private ownership. The entirety of these resources has not yet even been quantified. The existence of abundant natural resources which have not come under private ownership can be largely explained by the high transport and development costs resulting from the unique geo-ecological conditions of the region which, with currently available technologies, translate into low profitability and reduced capacity for guaranteeing the sustainability of agricultural activities.

The combination of these factors meant that for centuries Brazilian Amazonia remained an economic, demographic, and geopolitical frontier. However, since the mid-1960s, the dynamics of land occupation began to assume different features in terms of speed and rationale of the process. Lower transport costs resulting from government investment in the region's road network, tax breaks, and availability of credit for private investors—plus the emergence of nearby urban consumer markets (Brasília, Belém, Manaus, among others)—made it profitable to open up a range of agricultural and cattle ranching activities which were previously unviable in the region. The process of land occupation, as a result, has been one of unprecedented growth over the past three decades.

Between 1970 and 2000, for example, the region's road network doubled with the construction of over 80,000 kilometers of new roads.[10] To illustrate the impact of road development in the region, the maps below show the evolution of the total cost of transport per ton from each municipality of Amazonia to the capital of São Paulo state (proxy for the markets of the Center-South of Brazil) for the years 1980 and 1995.

---

10. Excluding the states of Maranhão and Tocantins, GEIPOT (2002)

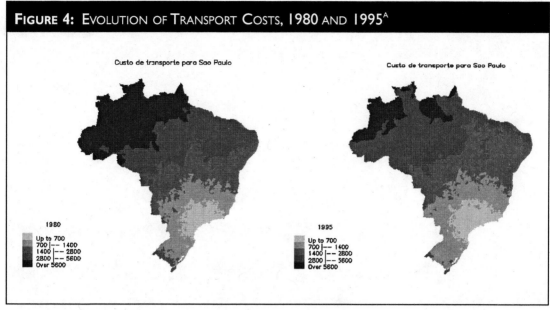

FIGURE 4: EVOLUTION OF TRANSPORT COSTS, 1980 AND 1995[a]

[a] Cost of one ton transported to the city of São Paulo, in 1995 Reals.
Source: Based on Castro (2000), drawn by IPEA.

The demographic and economic censuses for the years 1970 to 2000 contain a range of additional data about the dynamics of occupation of Brazil's Amazonian "frontier" region. In 1970, Brazilian Amazonia was a virtual demographic void, with an average population density of roughly 1.5 inhabitants/km², and less than 1 inhabitant/km² in the rural areas. A mere 12 percent of the region's territory was under private ownership, and over 80 percent of the area in those properties was not deforested.

In the following decades, the occupation dynamics of the "frontier" region accelerated rapidly. Population growth was mainly the result of migratory flows from other parts of the country. The population growth amounted to 3.5 percent a year—virtually double the national average for the same period. Nevertheless, that demographic density remains typical of Brazil's frontier regions: in 2000, the average density was of the order of 4.1 and 1.3 inhabitants/km2 for Amazonia's total and rural population, respectively. This indicates the rapid rate of urbanization of the region.

In economic terms, the dynamics of occupation resulted in a doubling of privately-owned land and the conversion of some 12 percent of the entire Amazonian region to cattle ranching, representing a total of 600,000 square kilometers.

The process of economic and demographic occupation of the Amazonian "frontier" was driven by, and dependent on, economic factors emanating from Brazil's Center-South and government policy. The low population density and a lack of economic infrastructure in the region meant that the price of land was significantly lower than in the rest of the country. This in itself was an incentive for integrating the Amazonian economy into the national economic mainstream. Integration took the form of private appropriation of land, with property rights often being acquired and consolidated by illegal means. It was followed by opening up the new land for agricultural and cattle ranching activities. Comparative advantages flowed from the relative abundance of cheap arable land and favorable climatic conditions.

At the same time, the introduction and expansion of cattle ranching (and other economic activities) demands and stimulates further immigration, hand in hand with calls for more substantial government involvement in terms of provision of basic services and infrastructure, including transport. This in turn lead to increased profitability of the livestock sector and underscored the rising competitiveness of the region.

Over the long term, however, the deciding factor for the continuing economic sustainability of agricultural and ranching activities has always been, and is likely to continue to be, the ability to adapt technologies to the geo-ecological conditions of the region. These adaptation initiatives are primarily funded by government. Over the past twenty years this has been the hallmark of both the soybean and livestock industries.

Geopolitical objectives related to sovereignty over the territory and the government's ability to control the economic potential inherent in the natural resources of the region have gone hand in hand with the strictly private economic ambitions associated with occupation and deforestation. In addition to the quest for access to markets in countries bordering the Amazonian region, aided by road access to the Pacific Ocean, geopolitical interests continue to be motivated by the existence of the long shared border with eight other South American countries. This makes monitoring and military protection of the Amazon region crucial, particularly when taking in to account that some of those countries play a major role in the international drug traffic.

Regardless of the importance of geopolitical factors in the expansion of the "frontier," once the cycle of government-led policies—aimed at constructing the main access roads across the region, attracting investments, and stimulating migration and productive activities—was over, the dynamic of the occupation process have became increasingly endogenous (that is, it has a life of its own). The economic activities acting as a driving force behind the opening-up of new areas for agricultural and livestock production and deforestation no longer depend on subsidies or government transfers from the rest of the country, even though the region remains dependent on the rest of the country for access to markets and finance.

The dynamics of occupation has not been homogeneous. The variety of geo-ecological conditions and the distance to the larger consumer markets, which imply substantial disparities between the costs of transport and economic exploitation, have resulted in a diversity of geographical, economic and social occupation dynamics. Several other non-economic factors also helped to bring about intra-regional variations in the actual history of land occupation.

In social terms, the agents of "frontier" occupation are differentiated by their particular motives, as well as by the resources available to them. Schneider's analytical outline (1995) asserts that the most distant regions attract only those "pioneering agents" who have lower opportunity costs. On the "consolidated frontier," on the other hand, one encounters a preponderance of more highly-capitalized agents with higher opportunity costs. These two types of agents act according to significantly different strategies.

The pioneers, many of whom are itinerants, base their strategies on speculation, and their economic activities largely depend on *nutrient mining*. Their predominant activities tend to be mineral extraction, logging, small scale agriculture and low intensity ranching, often adopted as part of a frontier survival strategy. The latter basically endows them with primitive property rights. The strategies pursued by these agents also vary in accordance with their perceptions regarding future prospects for a more "consolidated frontier"—in effect, the future willingness of the agents in the consolidated frontier to buy land (land profitability) and on the likelihood of infrastructure projects going ahead.

Government "control policies" have little real impact on the activities of such pioneers. The reasons for this are, on the one hand, the lack of an effective government presence in the frontier regions and the resulting inability to impose its will on the areas. A second reason is the more limited risk aversion by the pioneering agents who, simply put, are willing to do almost anything. These pioneering agents frequently co-exist, or are occasionally preceded by, small settlers (*colonos*), migrants, dispossessed farm laborers, etc., but this latter group's *raison d'être* is somewhat different. They devote themselves to cover their low opportunity costs through small-scale subsistence farming or by offering themselves as unskilled labor to larger enterprises. For the sake of convenience, we call the "speculative frontier" that part of the Amazonian region which contains such agents and *modus operandi*. The description is admittedly not entirely satisfactory, given that, even in this case, the process follows an economic logic. Speculative motives in the strictest sense of the word do not appear to be significant.

By contrast, the agents in the "consolidated" areas are much more geared to more commercial agricultural and, in particular, livestock production. Typical ranching production is on a large scale, with an increasing trend towards employment of new technology and improved pasture/animal management. Most of the current deforestation of Amazonia takes place in such areas. Deforestation caused by the larger landowners in these consolidated areas has less to do with the logic of "frontier occupation." It is more in keeping with a capitalist desire to invest in continuing expansion of business activities. Economic gains are the result of high productivity. Tax incentives, subsidies and land speculation are not particularly important for such agents. It is interesting to note that such agents regard the logging companies as playing a limited role in financing expansion of production and in opening new "penetration fronts". For the sake of convenience, we call that region where such agents and processes are in evidence the "consolidated frontier."[11]

Based on a distinction between the different types of frontier area, this study included three surveys to analyze the dynamics of the process. The first is devoted to an analysis of the microeconomics of cattle ranching in selected areas of the "consolidated frontier"—basically an examination of the strategies and economic motivation behind ranchers' propensity to occupy and clear forest and the potential profits arising from such activities. Chapter IV summarizes the main results of this research.

The second survey seeks to explain the activities of the agents in the "speculative frontier" and the logic behind them. It attempts to identify the main actors, to examine social relationships between them in those more remote regions, as well as their possible links to the agents operating in the "consolidated frontier." The next section is largely based on the lessons learned in the course of this field work.

The third survey consisted in the development of an econometric model that seeks to analyze the economic occupation of Amazonia as a systemic process. The model analyses some of the linkages between the main economic activities (agriculture, cattle ranching, and logging) from a temporal and spatial perspective. The principal results are summarized in the last section of the present chapter.

Expansion of the "Speculative Frontier" and the Deforestation Process

This part of the study is partly based on field research carried out between February and May 2002.[12] Its key objective was to better understand the motivations and social strategies of the actors operating on the speculative frontier. A special aim was to shed light on their natural resource appropriation strategies and to determine lines of succession regarding land occupation. The research sought: (i) to analyze how the principal actors (large and medium sized loggers and cattle ranchers, as well as small rural farmers) succeed in forging alliances; (ii) to obtain their opinions regarding the forest and its resources; and (iii) to identify the most common patterns of land appropriation.

The two specific areas researched were: (i) Novo Progresso and Castelo de Sonhos on highway BR-163 (Cuiabá-Santarém), forming part of the municipalities of Novo Progresso and

---

11. The terms are used for reasons of convenience. All "frontiers" are in principle speculative and cease to be frontiers once the process is complete. At the same time, the "consolidated frontier" is not in the strict sense of the word a "frontier" as such. But, insofar as the activities of the agents located there are *expansionist*, and in the specific case of Amazonia means that they secure benefit from conversion of forest to agricultural land, they are regarded as being at the very edge of this frontier. The term "consolidated frontier" is used to differentiate them from other agents with more *speculative* intentions. It is worth recalling that both "frontiers" and the whole range of processes involved are extremely dynamic—a continual flow of agents and strategies exists between the two extremes (speculative and consolidated), and the interests pursued therein are compatible and interdependent.

12. The full report (in Portuguese only)—Atores e Relações Sociais em Novas Fronteiras na Amazônia (Edna Castro et al, 2002) can be found in the World Bank's Web Site (www.bancomundial.org.br).

Altamira (Pará), respectively, and (ii) São Félix de Xingu, in particular the frontier regions towards the River Iriri and Terra do Meio,[13] also in the state of Pará. Expansion of the frontier in São Félix follows the massive occupation which took place during the 1970s in the area around Marabá, which spread outwards towards the south and southeast of Pará, where today municipalities such as Xinguara and Redenção exist. On BR-163, two fronts effectively meet—one pointing south from Altamira and the other going in a northerly direction from the border with Mato Grosso.

Interaction Between the Agents

The process of expansion and consolidation of the frontier in Amazonia does not come about in a uniform manner. Two general non-exclusive patterns can be identified: one where small agents make a first way in, and a second one where large agents clear and occupy new areas directly. In the past, in particular, public lands were mainly appropriated by small agents through colonization projects or "spontaneous" occupation. Capitalization of agriculture in the Brazilian South enabled small and medium settlers to sell their land there and purchase, for the same price, properties up to 15 times larger in Amazonia. In due course, these small agents, for a variety of reasons, ended up selling their plots to more highly capitalized agents. This first pattern of land occupation is thus marked by small agents preceding the larger ones.

The second pattern features deforestation and penetration directly by better capitalized agents—logging, mining and energy companies, large-scale cattle ranchers, etc. The smaller agents are still in evidence, but mainly as unskilled laborers serving their larger counterparts. As noted in the preceding chapter, this second group accounted for the bulk of deforestation activity and expansion into pioneering areas.

In this second model, the small agents are present in pioneer areas more closely linked to spaces opened up by logging companies or other large operators. The opening-up of new fronts in the forest is thus due to a marriage of interests between the smaller agents (rural laborers and dispossessed small farmers) and the logging companies. The latter need to make use of the scarce labor in the more isolated areas where timber is plentiful, where land is ownerless and where enforcement (of any kind) does not exist. Roads cut into the forest by logging companies provide access to timber and at the same time facilitate access by the smaller agents. The larger companies provide transport to lift out sick or injured people and account for a rudimentary cash circulation. For their part, workers are occasionally attracted by the promise of some day acquiring their own plots. They sometimes join teams of experienced loggers or simply decide to establish roots in these remote areas and devote themselves to subsistence farming based on "nutrient mining" in the forest. Depending on the degree of consolidation of the frontier and the location of their operations, the logging companies are substituted by large-scale ranchers who play a similar role. Similar patterns are followed in the case of mining companies.

A major difference exists between the two frontiers which are the subject of this research. Deforestation and the speed with which access ("penetration") roads are built are considerably less in the case of highway BR-163. The whole process in the area is now the purview of loggers and small farmers, with a number of cattle ranchers purchasing land with a view to future production, laying the basis for consolidating their property rights, speculating on the frontier being "opened up" (in other words, the paving of BR-163) and investing capital in more land to be eventually used for ranching. The arrival of sawmills is the harbinger of a process of major land appropriation devoted to cattle ranching and crop growing, as has already occurred in the states of Paraná and Mato Grosso. This process is accompanied by experiments with planting of rice and corn, and by the construction of drying facilities and grain silos in certain municipalities.

---

13. Region between the River Xingu to the east, state of Mato Grosso to the south, BR-163 to the west and BR-130 to the north.

In the case of São Félix, the agents of deforestation are mainly the large cattle ranchers.[14] Given the proximity of the cities of southern Pará (Xinguara and Redenção), the existence of slaughterhouses in those cities and the well-developed road network constructed and maintained through the combined efforts of logging companies, cattle ranchers and the local authorities, the physical and temporal distance separating the "speculative" and "consolidated" frontiers is substantially less. Land speculators (people specializing in selling land and consolidating titles) figuratively rub shoulders with rural smallholders, loggers (continually pressing further into the forest) and those agents more interested in production, in particular cattle ranchers with access to capital. In addition to timber extraction, local economies were initially boosted by mining activities (cassiterite and gold) and by *jaborandi* extraction.

The Land Market

The development of the land market in Brazilian Amazonia is a direct reflection of the deforestation process. The prospect of capital gain accruing from the purchase and sale of land means a run for land ownership—and land clearing is the main form of guaranteeing property rights. From the economic point of view, this process has its roots in the free-access nature of unoccupied, unclaimed land (whether *terra devoluta* or not; Almeida and Campari 1993 and Young 1998).

Traditionally, rising land prices in the south of Brazil (relative to the north) constituted an expulsion factor for migrants. As already mentioned, a small farmer in the south could easily double the size of his establishment by migrating to the north in 1970s. By the mid-1980s, he could obtain almost fifteen hectares in the north for one hectare in the south. This made existing land in the north increasingly inaccessible to poor landless farmers. At the same time, the process meant that less intensive activities such as cattle ranching were pushed northward to less expensive areas, further increasing pressure on the frontier. The by-product was an increase in deforestation.

An example is São Félix do Xingu. Land prices there are less than half those in the state of Goiás, from where many people emigrated to buy the initially cheaper land. Land prices in the chosen area start at US$3/ha, but eventually reach a level in line with neighboring markets—somewhere around US$300/ha. This means that the first occupants can eventually extract a substantial profit from occupation, clearing, land preparation, planting of pasture, acquisition of title and sale of the properties. The end-purchasers are typically medium-size to large landowners who pay equilibrium market prices for the land as in those adjacent properties. The price paid should thus reflect the rental value (*arrendamento*) of these lands, since they reflect better the true production potential.

The latter is an important consideration. Deforestation is frequently attributed to land price speculation. The landowners and cattle ranchers—the end-purchasers—who buy land already with legal title and ready for productive use, are prepared to pay a price per hectare which is equal to, or less than, the net present value of the activity for which they plan to use the land. *Grileiros* (land-grabbers) and speculators that invade unclaimed public land (*terra devoluta*), "tame" the land and accept a resale price at least equal to the marginal costs they have incurred, which are generally quite low. These agents, operating on the "speculative frontier," make decisions on land clearing and preparation on the basis of the certainty that their costs will be more than covered by the prices that ranchers are willing to pay, in short, that productivity of the land will exceed their own cost.

The marriage of interests is obvious, but the viability of the entire process is basically rooted in the cattle ranchers' willingness to pay. *The potential profit to be obtained from cattle ranching is ultimately the underlying motivating force, both for the initial agents and the final buyers, behind*

---

14. In the case of the Xingu, common knowledge is that most "felling" is done by landowners. According to a local leader "landowners are felling at a scandalous rate. IBAMA is concerned only with the loggers, but these have a different role in the deforestation process, opening up roads into the forest. The landowners' activities are a scandal, since they leave nothing alive on the ground: no trees, no animals. Landowners fell mercilessly and nobody cares to intervene to stop them".

the decision to clear forest and convert it to pasture. The whole sequence is largely dependent on this consideration. Without the profit incentive, interest in purchasing the cleared areas would be lower, resulting in slower rates of deforestation. The risk incurred by the pioneering agents on the "speculative frontier" is increased by the unpredictability of contestation of land ownership by the authorities (currently a minimal risk). This factor outweighs any uncertainties regarding future sale of the land they occupied and cleared (almost always illegally). While the agents of the speculative frontier run little risk of having their land confiscated, for the agents of the "consolidated frontier" this risk is largely non-existent.

In spite of the lack of accurate data, movements in land prices in the region underscore these hypotheses. Currently available supporting data apply to land in the consolidated frontier area, where the price trend is unequivocally downward-sloping (with the exception of a peak in 1986), while the rent/land price ratio is rising (Ferraz 2000). Both these indicators undermine the hypothesis that deforestation is driven by speculation. On the contrary, land transactions are based more on production potential than any other reason, and the price clearly reflects this. Buying "converted" land will tend increasingly to fall into the hands of capitalized agents whose primary objective is production. In the field interviews undertaken for the concept paper, this was the view expressed by the producers interviewed. The issue of pure speculation was considered less important. According to Wunder (2000, p.80) "…speculation may be a channel for short-run adjustment…, but it is not a cause of deforestation in its own right."

Grilagem *(Land Grabbing) and Consolidation of Property Rights*

Particular attention needs to be paid to what happens along the speculative frontier. This is the area where transformation of unclaimed native forest (*terras devolutas*) into properly titled and legalized establishments (in effect, with ensured property rights) for agriculture and ranching actually takes place. At all stages of this cycle, property rights are only assured through physical occupation of the land which has, initially, priority over any document proving ownership. Physical occupation encourages armies of land grabbers (*grileiros*) and squatters (*posseiros*) specializing in illegally taking over land and holding it until legal title is eventually established. Their activities are often financed by large logging and landowning companies. In this way, a kind of "private rule of law" is established to fill the vacuum created by frequent absence of the State.

Land grabbing (*grilagem*) is a key process of forest conversion to pasture. It is noteworthy that the high profits to be obtained from cattle ranching (discussed in the following chapter) are often due to the originally illegal appropriation of land which is camouflaged in subsequent financial returns. Land grabbing is not a new phenomenon. It is a feature of all the new fronts opened for ranching in Amazonia.[15] Countless formal complaints and lawsuits have been recorded by the Public Prosecutor's Office (*Ministério Público*) and by the Federal Police and IBAMA confirming the existence of *grilagem* in all of the regions studied (Treccani, 2001). A group of specialists derives its livelihood from such land transactions. They could be called "speculators" in the strictest sense of the word, but in fact a wide range of social actors is involved in illegally appropriating land and the many methods subsequently employed to sell it. The highest profits go to those who indulge in land grabbing and selling, and in particular to those who buy and sell plots in the more highly capitalized areas. Surprisingly, the landowners interviewed did not pay much heed to the possibility of their properties being confiscated by the State. People who acquire land in this way tend to establish themselves and build the necessary facilities, and often become successful entrepreneurs and eventually local and even regional power-brokers. They pay

---

15. The *grilagem* system obeys a certain *modus operandi*: (i) hired gunmen (*pistoleiros*) occupy and keep watch over the areas wanted by the cattle ranchers; (ii) the latter acquire false documents; (iii) they subsequently legalize their occupation of the land with ITERPA, the state land agency, which has been known to donate state land in these expansion areas, including Federal Government land passed on by the Agrarian Reform authority INCRA.

little notice to satellite monitoring as they employ new land clearing techniques not easily detected by remote sensing.[16]

Attention needs to be drawn to the role of INCRA in the land-grabbing process. Numerous lawsuits have been mounted against its employees. In this respect INCRA's activities involve a heavy dose of ambiguity. On the one hand, its mandate is to mediate and resolve conflicts. On the other, it is suspected of not carrying out its proper function of bringing order to the agrarian reform process. Whether there is truth in the many allegations made against it or not, when one considers the decisions taken about the future of land in the region over the past two decades, it is plausible to conclude that a well-organized land privatization policy, using small producers as a labor force, has been in operation for years.

Pursuit of such "policy" is also evident in the settlement projects undertaken by INCRA. Establishing unstructured settlements in isolated locations with no provision for public services, is the perfect mechanism for alienating the new settlers. Without schools, access to health care and the means to exploit their newly-acquired land, many settlers end up abandoning or selling their plots. In the words of a settler near Central: "...this is a land with no law, much humiliation and gangs of hired killers (*pistoleiros*)...life is worth nothing...murder is rife...justice non-existent." In all the settlements there are candidates-in-waiting on the edges of the properties. Having sold their plots, settlers head for nearby villages and towns where they buy urban plots and swell the pool of unskilled labor to be employed in the few economic activities in the region, or they migrate elsewhere. The outlook in general is one of increasing poverty and conflict.[17]

Finally, in addition to land grabbing, the conversion of forest to private properties with title *would not be possible without the "generous" fraudulent awards of title deeds and widespread corruption in the land market.* This subject was discussed with prosecutors from a number of State Public Prosecutors in the region. The officials acknowledged two key points in this respect: (i) that the above was a key consideration in the process of land occupation in Amazonia and (ii) that it was extraordinarily complex and difficult to handle. Public Prosecutors were at a loss as to how to undertake operations which could threaten established practices. Countless legal titles could be objected to in courts, since the award of title deeds recognized at official land registry offices for unclaimed public land requires a proper examination of previously registered titled ownership of such lands—an exercise which in the vast number of cases has never been carried out by the land registries.

---

16. Deforestation is taking place undetected by remote sensing. During the first year of deforestation smaller trees are felled and grass is planted simultaneously: a laborer distributes seed in areas while the mechanical earth movers "clean" the land. A year after the grass has taken root under the trees, cattle are introduced into the area. Thus livestock takes over the forest without, as far as the state is concerned, the trees being felled. The grass is burned during the second year as part of a secondary "cleaning" of the forest. Medium sized trees are felled, leaving only the larger ones. The grass grows again (its roots having survived the burning process) and this enables once again the cattle to graze on the spoiled area. Only in the third year does burning take place which destroys what remains of the initial forest cover, thus permitting detection by satellite. Following this model of deforestation, any action by the state is incapable of reversing the destruction that has already taken place or preventing the remaining forest from being ruined.

17. According to leaders of settlements in Tucumã and Ourilândia do Norte "without any incentive, the same will occur as always. I have managed to put up with this for 10 to 15 years. But to keep the producers here there has to be an environmentally-friendly project that should also be self-sustainable. We need technical assistance and financing to produce as a consortium. As far as we are concerned, the only way forward is to get along with the issue of the environment, because any solution that we try to achieve will have to be self-sustainable in order to avoid the land being abandoned time and time again".

Determinants of Land Occupation in Amazonia: An Econometric Model[18]

This study funded the development of an econometric model which analyzes the economic occupation of Amazonia as a systemic process. The aim was to provide an analysis (complementing those done on a micro level) of decisions affecting economic resource allocation and the relationship between the social actors presented in the foregoing section and in the next chapter.

The model analyses with a systemic perspective and on a regional scale the temporal and geographical patterns of the process of land occupation in Brazilian Amazonia by quantifying interactive relationships between the principal activities (timber extraction, cattle ranching, and farming) and their effects on the process of deforestation in the region, as well as taking into account the ecological and economic aspects which condition those relationships.

These models operate at the municipal level, based on data collected from demographic and economic censuses from 1970-1996, supplemented by geo-ecological information, together with data on transport costs obtained at the same level of geographic aggregation. To evaluate the importance of geo-ecological factors governing the land process, the municipalities of Legal Amazonia were aggregated into different categories according to precipitation, vegetation or relief.

The models employ panel data covering the whole period from 1970 to 1996. Since the number, area and geographical boundaries of the municipalities of Brazilian Amazonia underwent substantial changes over that period, the 1997 data for the 508 municipalities were aggregated into 256 "minimum comparable areas" (MCA) in order to produce consistent inter-temporal data sets and analyses.

Measurement of economic activities was done alternatively by geographic density of the product (relationship between the product and the area of the municipality) or by the percentage of the area of the municipality employed for the activity. The access conditions and transport costs are specified as exogenous variables (predetermined), based upon estimates of the total cost per ton transported from each municipality to the city of São Paulo in the years 1968, 1980 and 1995 (Castro, 2002).

The process of spatial interaction of the economic activities was specified in both dynamic and static terms. In the dynamic models, all regressors are predetermined, enabling an estimate to be made by the least squares estimation method which produced more reliable results. The static models presuppose that the spatial interaction between the municipalities is simultaneous, meaning that changes in the ranching activities in a given municipality can affect the others in the same period.

The complexity of the problem of estimating spatial and temporal interactions imposes restrictions on the number of variables that can be considered jointly. Therefore, three models were estimated: (i) the first analyses the spatial and temporal interactions of the densities of products in the three activities (cattle, timber and agriculture); (ii) the second does the same for the percentage of the area of the municipality employed in each of the activities; and, (iii) a mixed model analyses the spatial-temporal interactions between product density in cattle-ranching and crop farming with land use for all types of agricultural purposes (crops, pasture and fallow land).

Results

The model represents an original effort on account of its systematic treatment of the IBGE data at the municipal level and its systemic approach to the process, analyzing the temporal and geographic patterns of the process, bringing together logging, ranching and agricultural activities. This appears to be crucial in the understanding of the dynamics of the process and only very few studies to date seem to have succeeded in achieving this in the case of the Amazon (Angelsen et al. 2002). Nonetheless, while IBGE datasets are the only reliable source of information available for the proposed objectives, they present major limitations which call into question the empirical

---

18. This section summarizes the results of one of the studies contracted for this research—Moreira and Reis (2002). The complete text is available on the World Bank Website (www.bancomundial.org.br).

results of the models. The data are deficient in a number of ways, notably regarding timber production (measured only in terms of the agricultural establishments). The level of aggregation (municipal and even worse in the case of the MCA) also leads to imprecision of a number of measurements. In addition, the data give more prominence to the municipalities of Eastern Amazonia (Maranhão and Tocantins) than to the Central and Western areas of the region, precisely where recent deforestation is most intense. The gaps between censuses (five and even ten years) can also lead to loss of important temporal information.

At the same time, it must be recognized that (i) these are the best data available and (ii) few alternative models are capable of better representing the spatial-temporal process of deforestation in Amazonia. Since the data are not available in a better format, one must acknowledge that existing alternative analyses carried out at the same level of aggregation suffer from the same kind of deficiency. The lack of census tract data for previous years impedes carrying out panel data analyses which would be critical to cover the spatial and dynamic dimensions being attempted by the model. In short, the model would be a major contribution had better data been available; in their absence, the model's results loose credibility and must be interpreted with caution. This report thus not fully endorses its results, and recommends that a subsequent study be funded to allow its re-evaluation through the use of a more consistent and appropriate set of data, specifically census tract data for different years.

In spite of these shortcomings, it is worthwhile to present some of the model results. The objective is not only to analyze the results per se, but to discuss some of the key issues raised by the model, particularly the sequencing of activities and the importance of transport costs.

The key and most interesting results suggested by the model are:

- Both temporal and spatial interactions of the land occupation process of Brazilian Amazonia are significant, confirming the results of various similar studies (for example, Alves 2001, Chomitz and Thomas 2000, Fearnside 1993, Andersen et al. 2002, and Andersen and Reis 1997).
- In terms of precedence and causality, the model suggests that when the process of economic occupation is characterized by product density, timber extraction neither precedes nor is preceded by any other production indicators. This would suggest that, from a broad regional and long term perspective, logging is an autonomous process relative to the others. Mertens et al. (2001) report similar results.

    Although this result is perhaps due to a misspecification of the timber extraction process (which should be treated as an exhaustible rather than a renewable resource, such as agriculture), it is worth discussing it further. No other studies were found which formally demonstrate a dependency between logging and cattle ranching, where both occur, even though logging of course precedes ranching. It is a largely acknowledged fact that a long temporal gap exists between advance of the logging and the cattle frontiers, up to ten or even more years. No roads resist so many years inside the forest without intense conservation. Ranchers can follow previously explored areas and narrow the temporal gap behind the logging frontier, but they do not depend on the previous openings or roads. According to Westoby (Wunder 2000, p. 78), "... seeing logging as the main cause of deforestation in Brazil is a myth, given currency by those who would like a readily identifiable villain." Because this is not a crucial finding of this study, it is not important to reach absolute consensus on this difficult issue. It suffices to recognize that, while it is true that opening up of roads by loggers may make cattle ranching even more profitable, thus fueling deforestation, logging per se does not lead to major forest losses. As suggested in the concluding section, what is true is that disproportional attention is being paid to loggers relative to cattle ranchers: while logging should be closely monitored because it is unsustainable and largely illegal, loggers may not be as critical agents of deforestation as cattle ranchers, on whom increased attention should be focused;

- Rainfall is relevant in all the estimated models and thus would appear to be the principal geo-ecological conditioning factor in the occupation process. This outcome is widely accepted in the literature, and has been corroborated in recent work done by the World Bank (Schneider et al. 2000; Chomitz and Thomas 2000). The results of the estimates also suggest that the type of vegetation cover and relief are not particularly decisive geographical determinants in the occupation process.
- The impact of a reduction in transport costs in the two models suggests that, for three activities (cattle ranching, crop production, and timber extraction), it induces land use intensification and higher productivity. Increased productivity can lead to more or less deforestation depending on the price elasticity of demand. The only work that attempts to estimate such elasticity in the Amazon (Santana 2000) shows a somewhat counter-intuitive result for beef in Pará: demand is price inelastic (a result that can perhaps be explained by the recent incidence of foot-and-mouth disease in the region). This would lead to a paradoxical situation where a decrease in transport costs would lead to a decrease in deforestation.

This result runs counter to an unanimously accepted finding of the literature which holds that, while roads promote intensification and higher profits, they increase deforestation, implicitly assuming that the demand is elastic to price (Angelsen and Kaimowitz 2001, Cattaneo 2001, Pfaff 1997, Cropper et al. 1997, Andersen and Reis 1997, Laurance et al. 2001, Nepstad et al. 2000). The most recent work by Andersen et al. (2002) also runs counter to the usually more accepted finding. *This issue remains controversial, but there is no doubt that penetration roads in virgin forest lead to huge deforestation impacts.* In more densely occupied and consolidated areas this report remains agnostic.

CHAPTER 4

THE MICROECONOMICS OF BEEF CATTLE RANCHING IN AMAZONIA

As outlined in the Introduction, the observation that cattle ranching occupies virtually 80 percent of all the converted lands in Amazonia while appearing at the same time to deliver low rates of financial return, provides the main focal point for this study. The present chapter, perhaps the most important of the study, examines cattle ranching in the consolidated frontier, and in particular as practiced by the more "professional" and better capitalized ranchers. The objective is to assess ranching trends in the region, that is, the technical and economic viability of the activity and its prospects for sustainability as seen from the private producers' point of view. The main point at issue is that producers (ranchers) in the consolidated frontier effectively dictate current land occupation *trends* in the region: as long as cattle ranching remains competitive and economically viable, ranchers are willing to continue purchasing land from the earlier occupants of the speculative frontier. Regardless of the intermediate process which enables these "first" agents to profit from the whole cycle of occupation, conversion and acquisition of legal title to the land, the crucial point is that at the end of this process there is a productive activity (ranching) that can underwrite all the prior costs and expenses involved in its establishment—hence the need to study the microeconomics of cattle ranching in the region.

The aim here is not to analyze the "average" livestock economy in Amazonia, nor to describe the social aspects of the process. The latter will be examined in Chapter 5. It is worth recalling that the consolidated frontier largely coincides with those areas with intermediate precipitation indices (according to Chomitz and Thomas 2000, Schneider et al. 2000). Precipitation is between 1800 mm and 2200 mm per year, which are optimal climatic conditions for cattle production in Amazonia, according to this study. A far-from-coincidental relationship exists between "consolidation" of the frontier and appropriate climatic conditions for cattle production, which justifies the focus of the present research on the consolidated areas. The future expansion of the cattle ranching frontier, as predicted by the present research and based on the economic viability of the activity, will likely be restricted to the regions with similar climatic conditions, which reinforces the conclusions of the aforementioned references concerning the possible limiting role played by geo-ecological factors on the spread of deforestation in Brazilian Amazonia.

The Beef Cattle Economy in Amazonia: Brief Background

In the early 1970s, cattle ranching in the region was regarded as a predatory activity and as the principal cause of the spread of desertification. It eventually became a more profitable activity on account of the low land prices and subsidies of SUDAM (Amazon Development Agency). Furthermore, sales of timber extracted produced sufficient financial resources to cover the cost of the land and of clearing the forest, burning off residues, planting pasture and purchasing the basic cattle stock needed for primary herd development. For a few years after clearing, owners took advantage of the fertile soil conditions (resulting from the ash left over after burning) before abandoning the land (Tamer 1970). Schneider et al. (2000) draw attention to the fact that until the mid-1980s ranching failed to show a satisfactory financial return with traditional methods. Positive returns were only secured if fiscal incentives, speculative gains associated with selling land or favorable ratios of cattle to input prices were also obtained (World Bank 1991). Hecht et al. (1988) reported the same conclusion, and added overgrazing and low interest rates to the explanatory factors. Some of their simulations indicated that "modern" cattle ranching would only be viable under special conditions. It was postulated that large landowners were not really interested in incentives for cattle ranching per se, but had other considerations in mind—such as exemption from income tax, guarantee of property rights with the presence of cattle. In addition, standing forest was considered to be unproductive for purposes of the land tax and also for expropriation. Of course, the availability of financial incentives and subsidized credit for cattle ranching constituted an additional bonus.

In more recent research, Schneider (2000) draws attention to the low rates of return on cattle ranching, but is unable to discover any economic or financial justification for increasing herds. Arima and Uhl (1996) also point to low and even negative yields from beef production. However, their results also showed internal rates of return of over 20 percent when pasture management techniques were utilized. Similar returns had already been reported by Mueller (1977).

In one of the most comprehensive books on the subject, Faminow (1998) claims that many of the studies are inconsistent with observed practice. According to Faminow, the economic models are based, in general, on fixed technologies—an unsuitable approach to ranching in Amazonia. They almost always fail to take into account the variety of cattle production systems (dairy cattle, beef cattle, and multi-purpose cattle production), each of which involves a separate approach to breeding, different processing methods, different marketing techniques and a variety of investment, cost and turnover strategies. In addition, a range of factors not directly concerned with the economics of the activity or with the microeconomic aspects of livestock production were not included in the models which attempted to estimate "theoretical" rates of return. Many of these factors were in fact corroborated in the course of the brief field research for the *concept paper* as well as by the fieldwork carried out on the speculative frontier (see also Veiga et al. 2001)[19]:

---

19. It was clear from the research on the speculative frontier that cattle ranching is viewed by all local ranchers as a safe investment, as well as a financially viable one requiring little work. Mr. Alexandre Trevisan (*alias* Maneca) a rancher in Castelo de Sonhos, claimed that ranching was the "vocation" of the region: *"There is less money in ranching compared with logging, but it is much safer. You know where you are with it. I keep my feet on the ground and I know what I am doing and what I am capable of doing. Logging is different. If you get mixed up with problems out there (meaning problems with environmentalist movement, Ministry of the Environment, IBAMA, etc) or cannot lay your hands on the right documents to cut timber, everything comes to a grinding halt. Ranching is different. With a load of beef cattle you know you can get 10,000 Real for it or maybe more, you know that you owe your suppliers maybe three thousand, another thousand goes on gasoline, a few notes to pay your workers and you manage the rest. I reckon it's much safer"*. As far as the net return on the cattle frontier is concerned, the ranchers' representatives claim that this is fairly small, somewhere in the region of 5 percent. Even so, they see the business as attractive because it is safe, in comparison with investments in the formal financial system andin view of uncertainties regarding the government's economic policies. All ranchers fear having money confiscated and for them, cattle is a savings mechanism over which they themselves have control.

- cattle ranching is an obvious way of guaranteeing possession of land, one of the absolute top priorities on the "frontier";
- compared with agriculture, particularly seasonal crops, the risk inherent in ranching is extremely low in terms of market availability, sales and prices (despite a decreasing trend, meat prices have increased recently in relation to the main agricultural crops), climatic conditions and vulnerability to pests;
- unlike agriculture, cattle ranching calls for smaller upfront investments and provides quicker returns over a shorter period;
- cattle is an easily convertible, liquid asset;
- transport is relatively easy;
- ranching is not labor intensive;
- ranching is an optimal way of avoiding all types of enforcement (unlike crops);
- in the case of small producers, the activity provides indirect benefits. These include opportunities to produce other animal products, to produce manure for fertilizer and to have a stock of draft animals available—in addition to the gains to be made from selling timber (which also applies to larger producers);
- in the case of larger ranchers, the prospect of acquiring political and social status associated with being a large landowner/farmer in Brazil.

In an article dating from the same year, Young and Fausto (1998) made similar criticisms of the previous models and began to draw attention to the financial viability of cattle ranching in parts of Amazonia. Other studies indicated the feasibility of small-scale milk production (showing returns of 12 percent) and of beef cattle operations on reclaimed pasture land (returns of 12-21 percent).

Some of the criticism of the older literature for its assumption of low financial yields may have been due to new trends and practices not being reflected in older research. In fact, even today management and production techniques are still being tested and disseminated. Moreover, it could additionally be that the field data may have been collected in areas within the speculative frontier or elicited from smaller and/or less capitalized producers in the consolidated frontier, which in both cases would generate underestimates of the production potential. Given the ongoing expansion of cattle ranching in Amazonia, a whole range of contrasting results tends to emerge. These results need to be treated with even greater care when attempts are made to generalize to the whole of Amazonia on the basis of local surveys.[20]

In summary, *the economic analyses that suggest that cattle ranching has a low profitability in Amazonia or that it is only viable when benefiting from subsidies or speculative gains, are confronted by the inexorable advance of deforestation and by the expansion of the area dedicated to cattle ranching in the region. These analyses likewise fail to take proper account of the unquestionable reduction or elimination of government subsidies and credit facilities previously available to ranching in Amazonia.*

Schneider et al. (2000) acknowledge that "...the increase in the large scale and small-scale cattle and livestock herd continues, although a good empirical economic and financial explanation is lacking. A number of hypotheses, such as capital gains made on increased land values, require empirical verification."

---

20. According to Homma (1993), the expansion of the ranching frontier in Amazonia cannot be explained by profits to be secured from land speculation, since the risk of it being invaded by *posseiros*, its isolation from the main towns and the inexistence of special prices for agricultural products from the region have an impact on land prices. It is noteworthy that the price of an *arroba* (approximately 15kg) of beef in Amazonia is around 20 percent less than that obtained in the South-Southeast of the country.

Field Research[21]

In order to gain a better understanding of the microeconomics of beef production in Amazonia, field research was undertaken employing methodology proven by ESALQ (responsible for the work). This methodology consisted of an initial survey of data based on official government and regional sources, the drawing up of cost sheets, demarcation of study areas, and panel consultations and discussions with producers (fieldwork proper), followed by the estimation of cash flows based on the results of panel consultations and, finally, mathematical modeling.

The data, in addition to that obtained through interviews with the cattle ranchers themselves, was initially collected from SIDRA (IBGE Municipal Livestock Survey). When coupled with the maps of forest areas from PROARCO, it was possible to determine the municipal areas of potential interest for this study. The technical production coefficients were secured from EMBRAPA, INCRA, rural trade associations, consulting firms and retail outlets supplying agriculture/livestock-related items, as well as from local municipal governments and a number of other official bodies.

The production cost sheets were put together for the different types of systems pertaining to each area. They aimed to:

- provide information on production costs and herd yield rates for calculation of gross and net revenue;
- obtaining and storing numeric information on the usage frequency of inputs with a view to constructing a databank;
- obtain information on monthly cash flow over a 20 year period;
- obtain indices used in project evaluation—Internal Rate of Return [IRR] and Net Present Value [NPV].

The primary data were obtained on the basis of field research using the aforementioned panel methodology, in which face-to-face meetings were carried out with local producers. The figures obtained tended to reflect a consensus of the interviewed group, and mirrored regional reality fairly closely. Moreover, the same methodology was used in all panels by the same researchers, thereby minimizing data collection standardization problems.

Owing to the existence of different types of properties—physical area, size of herd, production system, level of technology available and different types of ranch management—it was not possible at a single meeting to secure information from each individual property. For this to be possible, it would have been necessary to submit individual questionnaires and to evaluate the results with statistical methods. Therefore, the research group was instructed to choose a single property which best represented others in a particular locality. In general, the properties indicated by most producers were of medium size and were neither particularly sophisticated nor particularly outmoded from a technical point of view. In other words, the properties fell into the "intermediate" range.

Once the representative properties had been selected, producers were questioned about their yields, and fixed and variable costs, including establishment costs, of their ranching activity.[22] All parameters were discussed by all producers until the numbers could be defined by consensus. Following field research, a number of parameters arrived at in the panels was compared with data from the IBGE Agricultural Census. The figures proved to be reliable and within the ranges observed in practice.

---

21. This and the following sections summarize the results of the work undertaken by ESALQ/USP contracted to examine the beef ranching sector in Amazonia—Barros et al (2002). The full version of the report (in Portuguese only) is available in the World Bank's Web Site (www.bancomundial.org.br).

22. The figures and costs claimed by participants were not related to their specific properties, but to the property which was pre-defined as being representative.

Selecting the Municipalities Involved in the Study and the Panels

Given the purpose of this study, the field surveys focused upon the most "consolidated" areas of the frontier, about which less information exists. The idea was to demonstrate the *potential* for cattle ranching in the region. This potential is already evident on those ranches with more experience. Further development is likely, as employment of better technologies by the latter is likely to be replicated throughout the region, given the need to remain competitive in this particular market. It needs to be emphasized that this work stopped short of reflecting practices which have to date been adopted throughout the whole region. By combining data about potential for growth in cattle ranching of the principal micro-regions and maps showing the forested areas of the main states (Pará, Mato Grosso, and Rondônia), the municipalities of Redenção, Santana do Araguaia and Paragominas (Pará), Alta Floresta (Mato Grosso), and Ji-Paraná (Rondôna) were selected for the field survey.

Altogether, eight panels were brought together with participation by 43 rural producers: four panels were set up in the state of Pará (Paragominas, 2; Redenção, 1; Santana do Araguaia, 1), one panel in Rondônia (Ji-Paraná) and a further one in Mato Grosso (Alta Floresta). The remaining two panels were assembled in the city of Tupã (state of São Paulo), one of the most important cattle producing areas in the country, with the aim to provide comparative results with the same methodology and by the same technical team to those in Amazonia.

Technical Parameters Adopted

In this and the following sub-section, the principle technical and economic parameters adopted in the panels are summarized, and a summary explanation of some of the more interesting or controversial parameters is presented. Table 10 indicates the size of the standard property and the use to which land was put in the six municipalities.

It can be seen that in Paragominas the properties are generally of large scale, with better maintained legal reserve areas. The majority of these reserves are not intact—most of the valuable timber has already been extracted. Since the Paragominas area has been occupied for a longer time and land ownership is more consolidated, agricultural activities have become more widespread, particularly over the past three years. The bulk of pastures was established 20 years ago and large tree trunks left over from clearing have already been burnt several times, or have disintegrated on account of the high temperature and humidity. This has led to lower windrowing costs and therefore larger-scale agricultural development. Moreover, pastures have deteriorated due to faulty management, overgrazing, soil compaction and low soil fertility in the area—all of which have caused a reduction in plant re-growth and a decline in grass productivity. Together with the flat relief and climatic conditions, these factors have favored the establishment of agriculture, particularly the growing of rice, corn and beans (soybean production still encounters difficulties at seeding time and during harvest, due to the high rainfall causing problems both at seeding time and during the harvest).

In the case of Santana do Araguaia, the area marked for forest reserve is zero, and in Redenção it is 50 percent. A visual inspection of the region reveals that the larger properties probably account for 50 percent of the reserves while smaller properties possess virtually no legal forest reserves. As in the case of Paragominas, grass productivity is also low, but in contrast to the latter the irregular relief and rocky soil have discouraged agricultural expansion in the region.

The colonization process in Alta Floresta is just over twenty years old. Government initiatives have led to easier access to land, resulting in the average property sizes smaller than for those in Pará. Legal reserve levels are approximately 50 percent, in keeping with the Forest Code.

Finally, the Ji-Paraná municipality has properties of smaller average size due to the way in which colonization was conducted in Rondonia. Plots were distributed along highway BR-364 in an organized but irrational manner, disregarding the limitations imposed by the relief and by the preponderance of poor soils. Thus, many of the deforested areas were not used to their full

TABLE 10: SIZE OF THE PROPERTIES AND LAND USE ADOPTED IN THE PANELS

Municipality	Size of property	Land use	Land Price (R$/ha)[a]
Paragominas (Pará)	Panel 1 12,000 hectares	50% - Reserve 50% - Pasture	
	Panel 2 15,000 hectares	60% - Reserve 36% - Pasture 4% - Agriculture	PF-R$ 300 PI-R$ 1,250
Redenção (Pará)	Panel 3 4,800 hectares	50% - Reserve 50% - Pasture	PF-R$ 300 PI-R$ 1,300
Santana do Araguaia (Pará)	Panel 4 3,200 hectares	100% - Pasture	PF-R$ 250 PI-R$ 2,000
Alta Floresta (Mato Grosso)	Panel 5 1,200 hectares	50% - Reserve 50% - Pasture	PF-R$ 250 PI-R$ 1,200
Ji-Paraná (Rondônia)	Panel 6 1,200 hectares	50% - Reserve 50% - Pasture	PF-R$ 200 PI-R$ 1,200
Ji-Paraná (Rondônia)	Panel 6 1,700 hectares	50% - Reserve 50% - Pasture	PF-R$ 200 PI-R$ 1,250
Tupã (São Paulo)	Panel 7 300 hectares	20% - Reserve 70% - Pasture 10% - Agriculture	PN-R$ 2,500
	Panel 8 300 hectares	20% - Reserve 70% - Pasture 10% - Agriculture	PI-R$ 3,300

[a] PF–Price of one hectare in areas that contain only forest.
PI–Price of one hectare in areas that possess a ranch with an established infrastructure for ranching.
PN–Price of one hectare of deforested areas and without an infrastructure for livestock (price refers to the state of São Paulo only, since the state contains no areas containing only forest).
Note: Land prices vary substantially, influenced by a number of different factors such as distance from the nearest cities, type of land, relief, the use to which land is to be put, extent of legally registered areas etc. The prices shown are approximate values referring to properties within a range of between 30 and 40 km from the nearest cities.

advantage, and some plots have in fact been abandoned. Table 11 below presents the main indices of animal production as indicated by each of the field research panels.

Comparing the indices, it can be observed that the best performance is reported by the Alta Floresta panel—for the rearing (*"cria"*), new-breeding (*"recria"*), and fattening (*"engorda"*) panels—and by Paragominas (for rearing and fattening). In the Paragominas region, efforts to develop and improve pasture management and reclamation techniques are noteworthy. The Alta Floresta region, a more recently colonized area, has well maintained pastures. Groups of well-organized ranchers operate in the area, interested in applying pasture management techniques. As mentioned previously, the technical parameters obtained from the panels were compared with those from the IBGE censuses. The comparison is summarized in Table 12 below.

TABLE 11: MAIN ANIMAL PRODUCTION INDICES AS INDICATED BY THE PANELS

Indices	Pará Paragominas		Pará S.Araguaia	Pará Redenção	M. Grosso Alta Floresta	Rondônia Ji-Paraná	São Paulo Tupã	
System type (a)	RBF	F	RF	RBF	RBF	RBF	RBF	F
AU (b) per hectare	0.71	1.53	0.86	0.83	1.18	0.91	1.14	1.08
Mortality rate (young animals)	4%	—	—	4%	3%	3%	2%	—
Mortality rate (adult animals)	2%	2%	2%	2%	1%	2%	1%	1%
Interval between births (months)	15	—	—	16	14	14	17	—
Pregnancy rates	75%	—	—	87%	88%	85%	75%	—
Daily weight gain in kg (annual average)	0.47	0.50	0.42	0.42	0.43	0.45	0.37	0.33

(a) System types:
   RBF: Rearing, new breeding and fattening (i.e., the property possesses matrices to rear its own calves as well as areas devoted to fattening and slaughtering cattle).
   RF: Property used for in rearing and fattening but needing to purchase calves
   F: Property solely engaged in fattening, needing to purchase calves or lean cattle
(b) AU = Animal Unit
Source: Data derived from field research.

TABLE 12: COMPARISON OF THE PARAMETERS OF THE PANELS WITH IBGE DATA, MUNICIPALITY OF PARAGOMINAS (PARÁ)

	Panel	IBGE
Size of Property (ha)	12,000	7,352
Mortality young animals	4%	10.2%
Mortality full-grown animals	2%	0.7%
Pregnancy rates	75%	82.8%
UA per hectare	0.7	3.0

As regards property sizes, it would not be possible to adjust the spreadsheet to other values since there is no way of evaluating the improvements ("*benfeitoria*") and machinery of a particular property of different size of that employed during field research. Any proportional reduction would not however affect income per hectare. With reference to mortality rates, it is relevant to note that the IBGE data refer to the 1996 Census, so the data were collected around 1994-95 The seven to eight years of difference relative to the data collected for this study are vital: mortality rate is one of the key parameters for ranchers, explaining much of the recent productivity gains with the adoption of techniques such as artificial insemination. Regarding animal units per hectare (AU), the panel data from the four remaining municipalities are almost wholly consistent with the data obtained from IBGE. In the case of Paragominas, it appears that IBGE data are flawed by a measuring error that overestimates the number of animals per hectare. If this parameter were altered, profitability would rise very substantially and would not entirely reflect reality; thus, the field research parameter was kept.

Climate

Among the environmental factors that most affect large-scale cattle ranching is, together with luminosity, the frequency and length of the dry season. These may be the main determinants of pasture growth and higher productivity.

The dry season is considerably longer, and rainfall is much lower, in the southeast of the country than in the areas closer to the Amazon. Together with high temperatures and high relative air humidity, these factors tend to reduce costs in the dry seasons. *This would appear to be the major factor favoring cattle ranching in the area, although it is important to emphasize that grass species used in the area are the same as those used in other areas of Brazil. Development of more appropriate grass species and varieties for this particular area could lead to better prevention of pests and invasive vegetation and, as a result, stimulate further rises in productivity.*

Production Costs: Pastures

The analyses begin with the rearing, new-breeding, and fattening system. To facilitate quick graphic visualization in a limited space, the cash flow over 20 years was divided into five parts, with each amounting to a 48 month period (four years). In addition, the columns were divided into seven main items. Since the proportion between the items is not substantially different in the four municipalities of the Amazon region, only one graph, that for Redenção, is presented below as an illustration.

Other Costs

The cost of purchasing heifers and cows is high only at the time that the ranches are established. The cost of machinery and farm implements is also relevant, but only in the case of Paragominas, where some properties choose to mix cattle ranching with other agricultural activities. As for livestock breeding, the purchase of mineral salts accounts for the overwhelming bulk of costs (about 75 percent) due to the high cost of road transport.

Other Systems

In the new-breeding-fattening system, the cost of acquiring the calves is by far the largest item. In the fattening system, this cost is even higher in view of the need to purchase heavier animals

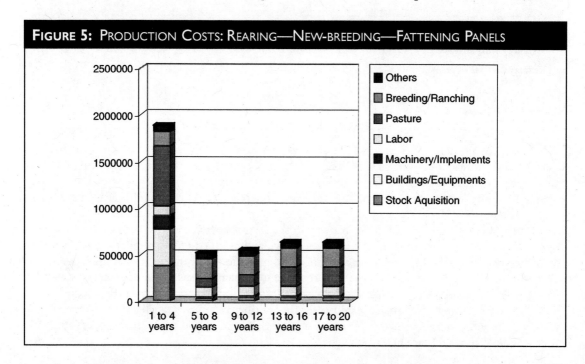

FIGURE 5: PRODUCTION COSTS: REARING—NEW-BREEDING—FATTENING PANELS

TABLE 13: NET INCOME PER HECTARE FROM CATTLE (IN REALS PER YEAR)	
Alta Floresta (Mato Grosso—Amazonia)	138.91
Ji-Paraná (Rondônia)	132.87
Paragominas (Pará)—complete cycle	95.39
Paragominas (Pará)	102.98
Redenção (Pará)	65.83
Santana do Araguaia (Pará)	95.80
Tupã (São Paulo)	65.32

exclusively for slaughter. The remaining costs associated with ranching vary, particularly grazing costs. Other costs tend to be in line with former systems (see Barros et al. 2002 for more details).

Net Income per Hectare

Net income per hectare is extremely important from the rural producer's point of view. It is the main factor analyzed at the time decisions are taken regarding which activities and which investments to be made. *The research showed surprising financial yields per hectare for virtually the whole range of activities studied in the Amazon region.* Soybean cultivation, still at an early stage, is highly profitable, although appropriate marketing structures still remain to be put in place. *In the case of cattle ranching, the high return per hectare is due to the high productivity levels obtained.* The dry season in the area lasts for approximately two months and the "moisture stress" of plants is much lower than that experienced in other parts of Brazil. Thus, while grazing animals in the *cerrado* or in the Southeast lose weight over a four-month period, this is not a problem in those parts of Amazonia studied.

In Mato Grosso, a *cerrado* area with widespread cattle ranching, productivity is low and the return hardly constitutes an incentive for producers. In reality, ranching in Mato Grosso is simply a way of occupying areas where landowners reckon that in the future agriculture might be a viable alternative—ranching simply fills a gap in the turnover of companies whose principal business interests revolve around agriculture.

In the state of Pará, and throughout Amazonia in general, very little historical or research data exist regarding the productive potential of soybean cultivation. During the field surveys, it was obvious that in Mato Grosso agricultural activity ends in Sinop. In the Alta Floresta region there is practically no agriculture at all, given the different relief environment (increased declivity).

Table 13 shows the average yields for 20-year long cattle ranching projects. It can be seen that net income per hectare in all areas of Amazonia is higher than that in Tupã/São Paulo, a cattle ranching area typical of Brazil's Center-South that acts as a kind of reference for assessing the performance of ranching sector in Brazil.

As far as the locations analyzed in Amazonia are concerned, yields can be regarded as "good to high." The price difference between the area under study and other parts of Brazil is remarkable. This is due to the fact that the areas studied in Amazonia, with the exception of Alta Floresta, fall within the area of "average risk" in terms of foot and mouth disease, which means that boned beef from the area is permitted in São Paulo but cannot be exported. In the state of Pará, live animal prices are 15 percent to 20 percent lower than in São Paulo. In Rondônia, the difference is of the order of 25 percent, but the current edaphic circumstances of that area translate into higher returns since pasture recovery is not required. The same applies to Alta Floresta.

Analysis of Yield

From the cost and revenue data collected in the panels, monthly cash flows were calculated for each area under study, which allowed calculating Internal Rates of Return (IRR). Constant prices

TABLE 14: INTERNAL RATES OF RETURN (IRR) FROM REARING, NEW-BREEDING AND FATTENING SYSTEMS

Location	IRR (land price disregarded) (%)	IRR (land price taken into account) (%)	Date of Panel
Paragominas	11.0	16.7	03/22/2002
Redenção	9.1	14.6	03/25/2002
Alta Floresta	14.5	15.2	05/21/2002
Ji-Paraná	11.5	N/A	05/15/2002
Tupã (São Paulo)	6.4	6.4	04/26/2002

Notes: Land prices in Amazonia: uncleared land—R$ 300/ha; cleared land—R$ 1,500/ha
Land prices in Tupã/São Paulo: both cleared and uncleared land: R$ 6,000/ha

TABLE 15: INTERNAL RATES OF RETURN (IRR) FROM NEW-BREEDING AND FATTENING SYSTEMS

Location	IRR (land price disregarded) (%)	IRR (land price taken into account) (%)	Date of Panel
Paragominas (a)	14.5	17.9	03/21/2002
S. do Araguaia	14.7	16.9	03/26/2002
Tupã (São Paulo)	3.8	3.8	04/26/2002

(a)—Fattening activities only
Note: Based on same land values as in Table 14.

of inputs and of the product (beef) were assumed. The price of beef is the cash price paid by local purchasers on the day that each panel was assembled.

As for land prices, for which data were collected in the course of field research, the following needs to be taken into account. If they were added to the investment costs, the total value of the property after twenty years would need to be included in the gross revenue of the final year, once the project is terminated. The estimate of the future price of land in each area will not be precise since numerous exogenous factors can affect particular tracts of land over time. In practice, however, it can be assumed that regardless of productivity, the price of land in 20 years will always be higher than land in virgin forest areas. From this, the conclusion can be drawn that inclusion of the land price would increase the financial yield on the activity. In the circumstances, it was decided to employ two alternatives: (i) in the first, the price of land is excluded, both in the beginning and at the end of the project period; (ii) in the second, the land price enters in both. In the first case, returns forms part of the calculations of both. Case (i) means that yields for beef ranching alone can be compared for the different regions, excluding the disparities which might arise from a rise or depreciation in the value of the land in question. *In all the subsequent analyses, the second scenario concerning inclusion of land price is disregarded.* The results obtained are summarized in the tables below.

Due to variations in the beef price, the internal rates of return were also calculated with the monthly averages of the historical prices in the four states involved in this analysis.[23] It can be concluded that the investments made in ranching activities in the Amazon forest region are higher, but *generate substantially higher levels of return than those derived from the same type of activity in the interior of São Paulo.*

It should be stressed moreover that the prices received by producers in Amazonia per *arroba* of fattened cattle in Amazonia are 20 percent lower on account of foot and mouth disease. As the disease is brought under control, producer prices can be expected to increase, thereby increasing the profitability of ranching across the whole region.

Regarding the circumstances under which land prices are taken into account, it should be noted that the price of land reflects the market value as stated by local ranchers themselves. Such prices reflect the process of land occupation and ownership largely through illegal acquisition, as discussed in the preceding chapter. They would therefore not be entirely appropriate in a social evaluation of cattle ranching.

Mathematical Modeling

Analysis of the panel results provides information on the returns to cattle ranching in Amazonia, and allows comparison with the case of São Paulo state. Ranching in the particular Amazonian areas studied competes with ranching in other parts of Brazil, but more importantly, it competes also with other possible activities within a property. The owner must thus decide whether to use his land for alternative agricultural activities or to leave it under forest cover until the right moment presents itself for clearing it for some activity connected with ranching. *In addition to the question of financial yield, the main factor that influences such decisions is the producers' perception of risk.*

Mathematical programming models permit to choose an optimal combination of economic activities on agricultural properties, taking into account or not the risks involved, as well as changes in the level of competing activities under a variety of scenarios.

Results of the Risk-free Model

In this model, the combination of activities that maximize profit is examined. This implies that the producer is indifferent to risk—his objective is to maximize income from the property regardless of the possibility of not being able to achieve required production or prices from his endeavors.

The model was developed for the area of Paragominas in Pará, since this was the only area in Amazonia's "arc of deforestation" where agricultural (crop) production structures is developing. In the other areas, ranching is the prinicipal activity. In Alta Floresta in Mato Grosso—a "transition" area between *cerrado* and rain forest—the undulating rocky soil is not so suitable for ranching, but has higher precipitation than the cerrado, and hence higher ranching potential.

The representative size of the model property is 15,000 hectare, of which only 4,500 hectares have been cleared. Equipment is financed with a credit from the Bank of Amazonia (with a three year grace period) at an interest rate of 8.75 percent. The available capital in the property is R$1.7 million. The net returns per hectare from activities carried out on this property, based on the field research data obtained in Paragominas, are presented in Table 16 below. These are high in comparison with other areas of Brazil, due to the production system employed in which already high-yielding crops need relatively small inputs. The value for soybean, in particular, is significantly higher than the one proposed by Costa (2000).

The results of the model indicate that producers would tend to employ the bulk of capital in the most profitable activity, in view of cash flow conditions. The results obtained indicate that

---

23. The price series were obtained from the CEPEA/ESALQ databank. A new internal rate of return was obtained for each of the existing prices. The set of IRRs regarding each average monthly value of the *arroba* in each region was represented in frequency histograms.

Table 16: Net Returns (in Reals per Hectare)

Rice	582.00
Beef Cattle	95.39
Soybean	517.21

Table 17: Absolute Deviations of Crops over the Past 10 Seasons (in January 2002 Reals per Hectare)

Rice	Corn	Soybean	Fat cattle
54.4	108.6	−177.5	1.2
50.4	125.9	139.3	−0.9
−25.1	16.5	241.5	−4.0
−45.0	−104.5	−15.8	2.7
−23.9	−111.5	55.4	6.9
−9.6	−65.7	−229.2	1.7
−45.4	−77.0	−97.9	−6.8
−146.0	−144.7	115.9	−4.0
140.9	33.6	−79.4	−1.6
42.9	53.4	47.3	1.8
5.9	165.6	47.0	3.0

approximately 1,600 hectares of soybean should be planted while the remainder of the land should be given over to forest.[24] *This result does not correspond to the observations of the field research, suggesting that producers are in fact not indifferent to risk.* In effect, producers are always seeking to manage risk through the use of additional sources of income, trading off part of the potential income for lower risk and diversifying activities on their properties.

Results of the Model Incorporating Risk

In order to incorporate the risk element in the model, deviations of gross revenue from a linear trend representing rural producers' expectations are used. In the proposed model, the deviations are obtained through simple linear regression where the dependent variable is the gross revenue and the explanatory variable is time. The deviations for each of the crops over the past 10 harvests are presented in Table 17. The values in the table are considered in the model which is now used with the purpose of minimizing risk, given a forecast income. As the forecast income declines, the deviations follow suit and the efficient frontier is obtained representing the set of points that can be obtained for different combinations of income and risk, given the available production factors.

In Figure 6, results for the property of Paragominas are shown. In Paragominas, agricultural activities are already established and a dynamic process of substituting ranching for production of annual crops is underway. The annual return on the property studied in Paragominas approximates

---

24. Cash flow restrictions prevent the area from being totally occupied by rice—in principle the most profitable crop on a *per hectare* basis.

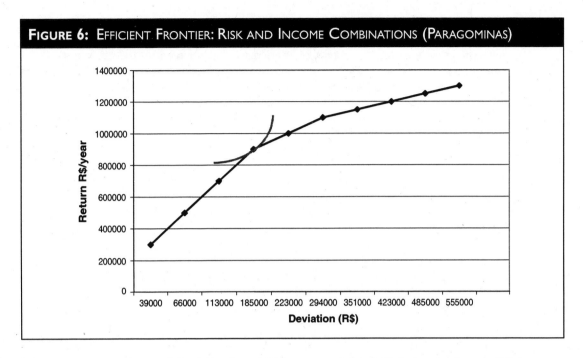

FIGURE 6: EFFICIENT FRONTIER: RISK AND INCOME COMBINATIONS (PARAGOMINAS)

R$900,000 on the basis of the following: 238 hectares of rice, 173 hectares of corn, 400 hectares of soybean, 4,924 hectares of pasture land, and 9,262 hectares of forest. Land allocation to the different agricultural activities, according to the results of the model, closely resembles actual practice in the field.

The model proved to be sensitive regarding area expansion in the event of price shocks. As the result of a price rise of one of the products, a redistribution of the area within the deforested area takes place first, and forest clearing may take place later. This can be explained by the fact that an increase in the crop or pasture area is conditional on making a substantial investment to clear forest, an investment which is rarely recouped from returns on the property.

At the point where the utility curve of the producer (red line) touches the efficient frontier, the point of equilibrium for the Amazon property is found. At that point, the producers' resources are allocated to 528 hectares of rice, 547 hectares of corn, 1,124 hectares of pastures, and 1,185 hectares of soybean. The remaining area, 11,613 ha, continues under forest. Such areas were obtained from the panel interviews—how farmers would allocate land between crops, pasture and forests (in percent terms) under the conditions being analyzed.

The similarity between model results and actual land use indicates that rural producers are not indifferent to risk. Ranchers or farmers do not have the means to open up large areas with the returns from current activities on their properties; revenues and cash flows fail to provide enough capital for short-term investments of this nature. This would suggest that any capital for expanding the cleared area must come from other sources.

Comparing the results obtained with the model and the field data, it is noted that relatively significant differences are present only in the case of soybean. Nevertheless, it is important to keep in mind the fact that soybean cultivation is rapidly rising and historic data regarding price and production for the state are only available for the period after 1997.

Some Simulations

The mathematical model permits simulations to analyze the probable impact of variations in certain parameters on the productive system. Such simulations help in the analysis of policies designed to optimize the system from the social point of view, and are available for this purpose. They are also helpful in understanding the ranchers' *own modus operandi*. However, only a limited range of simulations is presented here. The simulations were all done for the typical property described above.

The first simulation analyzed the impact of an increase of 10 percent in the net income from ranching, as could arise from an increase in the product price. In this case, the pasture area would increase by about 22 percent or 252 hectares, from 1,124 hectares to 1,376, of which 176 are from clearing forest area and the remainder from the agricultural area, with a significant reduction of the area under soybeans (59 hectares, representing about 5 percent of the original area of the crop). An increase of 20 percent (instead of 10 percent) in the net income from ranching displaces the agricultural area but does not alter the forest reserve area, because more benefit can be obtained from reducing the agricultural area in view of the high initial investment needed to clear forest.

With the aim of evaluating the amount of net return on the "forest" activity that would make it an attractive alternative to the others, simulations were made with different levels of annual net income from the forest area, starting at a base of R$10 per hectare. The simulation tested two situations:

- In the first, the producer takes into account not only the profit from the activities but also the risk involved. The actual conditions of the Paragominas property were thus reproduced, with a yearly income of some R$900,000 (the point at which the profit curve touches the efficient frontier). The results indicate that a net return of more than R$45.00 per hectare makes it worthwhile for producers to retain forest at the expense of the other agricultural activities considered;
- In the second case, the producer is indifferent to risk, selecting activities from a profit-maximizing viewpoint. In this case, a net income of around R$200 per hectare would be needed for the "forest" activity to take priority over the others considered in the study.

These two results show the opportunity costs involved in maintaining a forest reserve on a private property, with or without taking risk into account. The low value of the "risk" case is surprising, as producers would be willing to earn a relatively low net return for leaving part of their properties under native forest. This gives an idea of the magnitude of the compensation eventually required and the costs of implementation of such policy in the region. We return to this issue in Chapter 5.

Finally, simulations were made regarding the effect of a tax on deforestation as a way of obliging ranchers to internalize at least part of the social costs of deforestation. Two taxes were considered: one of US$15 per hectare and another of US$20 per hectare (both based on an exchange rate of R$3.55 to the US dollar). Table 18 below presents the results of simulations which apply this hypothetical tax on a typical farm in Paragominas.

TABLE 18: RESULTS OF DEFORESTATION TAX SIMULATIONS

Activity	Original areas of model (ha)(a)	New areas (ha) with US$20/ha tax	New areas (ha) with US$15/ha tax
Rice	239	1,055	729
Corn	173	176	267
Soybean	401	476	439
Fat cattle	4,925	2,666	3,329
Forest	9,262	10,627	10,236

(a) —distribution of the activities of the model based on an annual net income of R$ 900,000

The first column shows the types of activity under consideration. The second represents the area occupied by each of the activities in the original model: a farm property covering 15,000 hectares, with R$1.7 million in available capital and with an income of R$900,000 optimized in terms of risk and return. The third and fourth columns indicate the new optimal values after the imposition of a deforestation tax.

The tax has the effect of leaving a larger area under forest: about 1,365 hectares larger with the US$20 tax, or 974 hectares with the US$15 tax. Considering the second case, it is seen that the area under pasture is 1,596 hectare less than in the base case, which is 63 percent more than the increase in forest area. The remainder of the reduction in pasture (622 = 1,596 less 974) is used in the following way: 490 hectares for rice, 94 hectares for corn, and 38 hectares for soybean growing. In other words, in addition to inducing preservation of the forest, the tax led substitution of cattle ranching by cultivation of corn, soybean and especially rice. This in effect means that owners would intensify land use, something to be expected if they face higher land prices.[25]

As indicated earlier many additional simulations would be possible but are beyond the scope of this study.[26] The interested reader is referred to Barros et al. (2002).

Final Considerations and Trends

Experience to date points to the difficulties but also to major advancements of ranching in the region. Despite the problems, many lessons are being learned in terms of cattle and pasture management technology. Intensification and specialization can bring high returns, but they demand a long time span. Cattle ranchers are likely still at a relatively early point on the learning curve, with current figures reflecting no more than short-period trends. They are rapidly becoming "professional" as markets become increasingly competitive, as seen in the ongoing trend towards intensification and increased production efficiency. The experience of the American West would appear to hold lessons for what is occurring in Amazonia—*initial economic failure does not impede expansion of the "frontier," but rather speeds up the process of adapting to new managerial and technical methods.*

The growing perception by local cattle ranchers of the potential economic returns of livestock in Amazonia has at times involved high investment in pasture improvement on reclaimed land: over 600,000 hectares of abandoned land were reclaimed through planting of improved pasture varieties, at an approximate cost of US$260 per hectare, permitting densities of 1-1.5 heads per hectare and generating returns on investment of up to 13-14 percent. This experience will probably become the norm in the Paragominas area and in others where productivity is beginning to decline.

Generally speaking, the high economic returns to ranching in the area are due to the availability of good pastures as a result of high precipitation (intensity and frequency), high average temperatures, high relative air humidity and of the type of grasses used. This suggests that livestock

---

25. It is interesting to compare the amount of money required as a compensation for farmers not to deforest a marginal hectare of forest (R$ 45 with risk, R$ 200 without risk)—a subsidy—with the value of a tax which would make farmers reduce deforestation as presented in the last simulation. In the first case the value reflects the opportunity cost of maintaining forests on a private property (it was obtained by calculating the value of forests that would make producers indifferent between such activity and agriculture or ranching). The second has been calculated applying a tax on every hectare deforested (an ad hoc increase in costs applied to every hectare deforested). The simulations suggest both a tax and a subsidy of R$ 45/ha (approximately US$ 15/ha) would make the typical producer reduce the deforested area by about 1,000 ha in the property analyzed, corresponding roughly to 17 percent of the deforested area. The obvious fundamental difference between the two policies is who bears the costs.

26. One of these simulations might look at the opportunity cost of capital. The value adopted in the model's basic case scenario (8.75 percent) is typical of the more capitalized farming system in Brazil according to CEPEA/ESALQ. But cattle ranching is an activity which essentially does not operate with credits in Brazil, so that even the value adopted might be an unrealistic overestimate of the real capital cost faced by ranchers.

and pastures in regions without the same geo-ecological conditions may not reach the same productivity, as in the case of the high rainfall zones already mentioned.

Looking forward, a number of factors could either foster or militate against profitability and expansion of cattle ranching in the area (see Veiga and Tourrand, 2001).

Factors that Could Reduce the Performance of Cattle Ranching
Problems associated with pasture expansion
- Large scale use of *B. Brizanta* to replace the old *colonião* pastures. The former adapts very well to the soils and climate of the Amazon region but may lead to the emergence of contiguous areas of monoculture, and thus to the spread of diseases and pests that significantly reduce the production of biomass, and to higher maintenance costs.
- The formation of large air corridors due to the lack of natural barriers previously provided by high forest could decrease relative humidity and increase the propensity of the pasture to dry out, reducing its regenerative properties during the dry season.
- An increase of abandoned areas and incorrect pasture management favors the spread of invasive plants, which compete with the different varieties of cultivated pasture. The problem is particularly serious in the areas of Redenção and Santana do Araguaia. The use of herbicides is problematic because the active ingredients killing invasive plants have a deleterious effect on the cultivated grass.

Problems associated with expansion of the cattle stock
The costs currently involved in fighting diseases and parasitic agents are not particularly significant today. However the expansion of the areas under pasture and the trend towards establishing adjoining properties facilitates direct contact between the herds on neighboring ranches. This in turn can lead to diseases and parasitic agents affecting a wider area and to increased production costs.

Enforcement
The bulk of producers consulted in the course of the field research were aware of the need to retain 80 percent of the original forest cover as specified in current law (*Medida Provisória 2166*). They also knew of the risks associated with deforestation. Corruption was cited on a number of occasions as the key factor which encouraged flouting of the law. Nevertheless, the existence of the law and a fear of committing infractions help to curb even wider deforestation.

Factors that can Increase the Performance of Cattle Ranching
The growing herd size and local development have led to the setting up of organizations to protect local interests and promote economies of scale. In all areas researched, the existence of municipal producer unions and active state organizations was noted. They focused on the control of foot and mouth disease, organizing fairs and other events aimed at stimulating trade and on training courses and lectures geared to the improvement of technology and introduction of new ones.[27]

As the areas develop and foot and mouth disease is brought under better control, it is probable that other investors will be attracted and that industrial meat cooling plants from the south and southeast of Brazil might relocate to the area. This will no doubt lead to price increases and larger markets for the product.

The latter is one of the potentially more favorable results of expanded cattle ranching in Amazonia: the opportunities offered by domestic and external markets. According to Santana

---

27. In Alta Floresta, for example, the high costs involved in purchasing mineral salts led producers to set up a company to manufacture this particular input within the municipality in order to reduce costs. At present, the factory is investing in research with a view to improving food supplement products.

(2000), this prospect favors development of the beef cattle industry, so long as product quality is a priority. However, development of the export potential for beef is closely linked not only to international demand but also to the international price system, a wide range of sanitary restrictions and to protective measures in the importing countries (Haan et al. 2001).

In the same way in which producer organizations formed rapidly to deal with foot and mouth disease, there is also the prospect of the larger ranches in the area adopting mechanisms to track and obtain younger stock, without the need to use hormonal products or to confine cattle.

The future of the cattle ranching economy will also largely depend on the future cost of transport and on the incorporation and consolidation of technologies most suited to the region. It will, of course, also depend on the relationship between ranching, crop production and timber extraction, on changes in markets (particularly urban growth in Amazonia), and on the opportunity costs of opening up new areas (as opposed to the option of intensification). Of these, transport cost and technology are key factors. New production technologies will certainly arise from a combination of innovations undertaken by ranchers and research results obtained by EMBRAPA and other agencies. Transport costs will depend on government investment in the necessary infrastructure, especially roads—as foreseen in the *Avança Brasil* Program. The effects of increased urbanization of the Amazon region on cattle ranching are largely unknown and studies on the subject are still in short supply.

In conclusion, the key question is the following: will new areas still need to be cleared even if cattle ranching uses existing land more intensively? *Our view is that the trend towards accelerating growth of the cattle stock and of the area under pasture will continue.* Pressure for expansion of the "cattle frontier" will result not only from the dynamics of the ranching business itself (increasingly consolidated and profitable, with increased opportunities in the local markets as well as those in the Brazilian south and abroad) but also from pressure on the "agricultural frontier." Recent studies undertaken by the Bank in partnership with IMAZON suggest that there are natural barriers in the way of expansion of both the "ranching" and "agricultural" frontiers. The frontier features extremely high precipitation levels in the areas close to the "heart" of the dense rainforest. Experience of the *bragantina* area of the state of Pará offers irrefutable evidence that only very few, if any, economic activities are possible in such areas, and that perhaps only logging may make economic sense. The question is whether lessons have been learned and disseminated, or whether further deforestation and land use conversion will take place before it is realized that such areas are unviable for ranching. There is thus an evident need to implement effective zoning to bring order to land use. We return to the issue in the final chapter.

CHAPTER 5

SOCIAL COSTS AND BENEFITS OF DEFORESTATION

In the preceding chapter, the economic viability of cattle ranching in Amazonia was discussed, based on studies carried out in representative areas of the "consolidated frontier." The results suggest that in some areas relatively high rates of return are possible, although this is not the case for the whole area. The aim was to demonstrate that a trend towards more "professionalization" and more widespread use of improved ranching technology is the key to explaining the dynamics of deforestation in Amazonia. The driving force behind the ranching business is the real prospect of profit from the activity. This in turn motivates the decisions made by the chain of agents involved, from the primary speculators at the beginning of the process to the capitalized professional entrepreneurs on the consolidated frontier. A wide range of intermediaries is also involved.

The private profit to be obtained from cattle ranching does not ensure that wider social benefits necessarily flow from the activity. It is necessary to observe the social and environmental effects (in the widest possible sense of the term) of the activity. From the social standpoint, an assessment of cattle ranching must incorporate all the potential social benefits associated with it, whether they are local (such as generation of employment, income growth, better living conditions for the population, access to services etc.) or national (such as lower beef prices, higher exports, and increased protein intake by the poorer population). From the sum of social and private benefits obtained from ranching, some idea of the total potential social gains associated with the ranching business and deforestation can be obtained.

On the other hand, the negative environmental, social and cultural effects of clearing and ranching must also be measured—a range of factors which could be regarded as "social costs." These costs of the activity can then be compared with the "social gains," in an effort to arrive at a cost-benefit evaluation of cattle ranching and associated deforestation.

The social evaluation must take into account the opportunity costs of cattle ranching. In other words, even if private gains exceed the respective costs involved, it is necessary to examine alternative activities to ranching which may be able to compete on the same scale as ranching, such as forest management. In short, it will be necessary to compare the net social benefits of the two activities and not *accept cattle ranching simply on account of its viability in terms of its potential to generate private profit.*

A social evaluation of deforestation must cover at least three different levels: (i) local society (inhabitants of the region), as well as the wider perspectives of the (ii) national and (iii) global population. Each of these perspectives may lead to differing results in terms of, for example, admissible amounts of deforestation, the locality in which deforestation takes place, etc.

This simplified framework presupposes that risk-free decisions are taken, but this is, of course, is far from being the case in the real world. Any assessment of the social costs of deforestation will always be susceptible to a fair degree of uncertainty—even if the analyses point to a net social benefit it can be argued that the imprecision inherent in analyses of this nature (for example, the intrinsic value arising from the irreversible loss of certain species of flora or fauna, or the possible effects of deforestation on micro-climates or on rainfall patterns) call for broader criteria than strict economic cost-benefit assessment.

This study does not claim to provide answers to all the questions posed and at all possible levels of analysis. This chapter makes a rough assessment of the social costs and benefits linked to the spread of cattle ranching. Next section briefly examines the role of fiscal incentives in the expansion of cattle ranching and deforestation. The issue is discussed whether ranching is profitable only as the result of government incentives. Section 2 deals with estimates of the economic (environmental) costs of deforestation made for this study. Finally, Section 3 presents some economic indicators on alternative activities, summarizing the various estimates that might serve as a preliminary basis for a social cost-benefit analysis of deforestation.

As already mentioned in chapter 2, the Annex to this report presents the evolution of some socioeconomic indicators in Amazonia. They are not included in this chapter because it is not possible to directly attribute the observed gains to deforestation. This study adopts a conservative approach with regard to the possible size of local socioeconomic benefits by not considering them as effects of deforestation.

Government Incentives: Subsidies as a Basis for Cattle Ranching and Deforestation?[28]

The contribution made by government credit and other fiscal incentives to deforestation remains a controversial issue. Until the 1980s, it was generally accepted that ranching basically served as a mechanism to guarantee land ownership, for enjoying the benefits flowing from official credit and subsidies and as a way of making speculative gains (Hecht 1993, Tardin et al. 1982). From this point of view, fiscal incentives played a decisive role in the expansion of cattle ranching in Amazonia during the 1970s and 1980s for the simple reason that large-scale ranching was deemed not to be economically viable without fiscal incentives (Hecht 1985, Hecht et al. 1988, Browder 1988). Incentives were therefore critical to explain deforestation in the region (Binswanger 1989, Mahar 1989).

Schneider (1995) claims that agriculture in Amazonia received incentives of the order of US$3.172 million (1990 Dollars) between 1971 and 1987, representing an average input of US$300 million per year. In a seminal article on the role of fiscal incentives in Amazonia, Yokomizo (1989) showed that cattle ranching projects represented 58 percent of the total number of projects approved by SUDAM in 1989. Most of these were in Pará and Mato Grosso—39 percent and 25 percent of all agriculture-based projects, respectively. The bulk of these projects were in the cattle ranching sub-sector. Examining the data on deforestation in these two states for the same year, it was estimated that FINAM-supported projects were responsible for approximately 16 percent of deforestation after 1970 (21 percent in Mato Grosso and 7.5 percent in Pará).

---

28. This sub-section draws largely on Pacheco (2002b).

These estimates assume that the total area benefited by these projects was actually converted to ranching purposes. It is worth noting however that in the vast majority of cases this was not the case (Hecht 1982).

Ranching projects tend to be long-term undertakings, generating little employment (an average of only 36 jobs per project) and showed the lowest rates of financial return compared with other sectors receiving incentives. Given the dearth of information, it is not possible to give a more precise estimate of the area which was actually converted to pasture under the auspices of each project. It is also not possible to determine whether the conversion to pasture was in line with the original officially-approved schemes (SUDAM 1995).

In the early 1990s, subsidized credit—viewed as the driving force behind ranching or deforestation expansion—was terminated, and the various tax breaks and credit-based incentives were gradually phased out. More rigorous monitoring was also introduced. There are, however, few signs of a reduction of deforestation rates over the same period. In fact, areas with medium-sized properties actually showed very high rates of deforestation, although they were the beneficiaries of only limited incentives provided by SUDAM. As indicated by a number of economic studies, a range of other factors may be behind this process (Faminow 1988, Schneider 1995). The basic explanation is that cattle ranching tends to expand in Amazonia for the simple reason that it is a low-risk and profitable enterprise for those involved (private gain) and that ranchers have more in common with capitalist entrepreneurs whose main objective is to continually expand their businesses in order to maximize profits rather than to await the direct or indirect benefits provided by government or to engage in pure speculation (Margulis 2001).

Table 19 below shows that from 1991 to 1999, projects in the agricultural sector represented the majority of approved projects, but accounted for only 16 percent of total FINAM funds (US$588 million, or US$64.4 million per year). Since 1996, incentives have primarily been allocated to the industrial sector (Arima 2000).

Schneider (1995) makes an estimate of the probable effects of fiscal incentives on the expansion of ranching activities in Amazonia for the years 1980 and 1985 (Agricultural Census years) assuming that the available incentives benefited the largest ranchers. He concluded that FINAM incentives accounted for 17 percent of the cattle stock in 1980 and for 25 percent in 1985 (on 0.2 percent and 0.4 percent of the properties respectively). He also claims that since the 1980s the most dynamic sector of the cattle ranching industry is accounted for by properties with fewer

TABLE 19: FINAM FISCAL INCENTIVES BY SECTOR (IN 2000 US$ MILLION)

Year	Agribusiness	Agriculture	Industry	Services	Total
1991	57	41	317	100	514
1992	38	100	249	64	452
1993	4	73	27	6	110
1994	52	12	31	117	213
1995	8	17	44	25	93
1996	47	26	47	56	176
1997	84	92	181	26	384
1998	121	158	443	39	761
1999	146	61	600	6	813
Total	555	580	1,940	439	3,515
No. of projects	125	319	235	54	733

Source: SUDAM apud. Arima (2000), adapted by Pacheco (2002b).

TABLE 20: MAXIMUM HERD NUMBERS BENEFITING FROM FINAM INCENTIVES [A]

	Establishments	Herd
1000 units, 1980	54.8	3,989.1
1000 units, 1985	90.2	5,358.7
1000 units, 1995/96	148.9	12,058.5
% with FINAM Incentives, 1980	0.2	16.5
Idem, 1985	0.4	25.4
Idem, 1995/96	0.1	7.5

[a] Includes the states of Acre, Amapá, Rondônia, Amazonas, Pará, and Roraima.
Source: 1980 and 1985 based on Schneider (1995), Pacheco (2000b) using IBGE Agricultural Census for 1995/1996 and SUDAM data. See Schneider (1995) for methodology.

TABLE 21: RURAL CREDIT FOR NORTHERN REGION (IN 2000 US$ MILLION)

	1998	1999	2000	2001
FINAM livestock	158	61	57	47
Total rural credit (a)	114	195	284	327
Total of rural FNO	67	137	197	115
Livestock total	50	83	167	168
FNO livestock	29	62	114	79
Rural FNO/rural total (%)	58.4	70.3	69.4	35.2
FNO livestock/total livestock (%)	58.0	74.1	68.4	47.2
FNO livestock/FNO rural (%)	43.1	45.0	57.9	68.7

(a) Credit for the states of Acre, Amapá, Amazonas, Pará, Rondônia and Roraima)
Source: Adapted from BCB, *Rural Credit Statistical Yearbooks, 1998-2002*.

than 100 animals, and therefore not in receipt of incentives. Using the same methodology, Pacheco (2002b) estimates that in 1996 ranching projects funded by FINAM benefited 0.1 percent of the properties and about 7.5 percent of the cattle (Table 20).

Other studies show that the majority of ranchers, particularly those of small and medium size, do not benefit and have not benefited from fiscal incentives (Browder 1988, Arima and Uhl 1997). The latter authors suggest that only 4 out of a total of 47 ranchers interviewed received any kind of fiscal incentive and none of them claimed that they were operating in Amazonia in the expectation of obtaining fiscal incentives.

In addition to the incentives provided by FINAM, two other instruments have played an important role in financing agriculture and cattle ranching Amazonia: rural credit and FNO funds. Table 21 below shows that rural credit allocated to the northern states of Legal Amazonia is on the increase, although such finance still only accounts for between 2 and 3 percent of the country's total rural credit provision (Gasques 2001). The allocation attributed to FNO varied from 35 to 70 percent of the total amount of rural credit and the proportion offered by FNO to the ranching sector has been significant, increasing from 43 percent in 1998 to 68 percent in 2001.

It is difficult to measure the effects of the FNO on the dynamics of cattle expansion in Amazonia. However, this has evidently made a modest contribution in terms of the region as a whole and the FNO certainly played an important role in increasing the herds of small producers with limited access to investment credit. Cattle purchased with FNO funding (one million heads) represented an increase of 9 percent in the total bought in the years 1990-2001: 35 percent of the rural credit under FNO control was allocated to ranching activities over the entire period. More recently such participation has been increasing.

The social effects of the FNO contribution were mixed. The preferential credit allocated by the FNO had a positive influence on expanding the herds of small and medium-sized ranchers, although the main aims of product diversification and increased family income have not been fully achieved. Many small farmers encountered difficulties in turning to milk production as their main activity, in part due to problems with the FNO disbursements procedures (Andrae and Pingel, 2001) and due to the farmers' inability to develop a milk processing and marketing infrastructure which would have facilitated contact between producers and the emerging markets for dairy products (Veiga et al. 2001). Nevertheless, producers with more experience in cattle management succeeded in benefiting substantially from the program.

In short, fiscal incentives played an important indirect role in the past in building up the infrastructure and production basis for cattle production activities. Moreover, given the general unfamiliarity with cattle production techniques in Amazonia, it could be claimed that fiscal incentives helped considerably in the initial difficult stages of production involving a great deal of trial and error, particularly back in the 1970s (although much of the money made available was not in fact actually spent on livestock production). As modern techniques came on stream over the years, cattle production became competitive and FINAM funding for ranching was drastically cut back. At present, in view of the more rigorous accounting controls over resource allocations, fiscal incentives do not rank high on the list of factors which could explain the observed profitability of the sector and deforestation in Amazonia.

Estimate of Economic (Social) Cost of Deforestation in Amazonia[29]

With the aim of estimating the economic costs of deforestation in Amazonia in monetary terms and to enable a quantified comparison with the benefits, a limited environmental valuation was carried out. In the case of deforestation, this involved identifying the values from the future stock of forest which ultimately determines the future scarcity of "lost" environmental resources (i.e., their future value).

In theory, it would be necessary to include dynamic parameters to determine the economic cost curve of deforestation over time rather than using a value for this cost at any particular moment in time. However, identifying such curve would be a highly unreliable exercise and future values would need to be discounted over time—in other words calculated on the basis of present values, requiring the use of a social discount rate. Determining this rate would be difficult since it implies estimating the rate at which present consumption would need to be exchanged for future consumption without the possibility of consulting future generations. Thus, scenarios for these future economic values and for different discount rates were adopted in order to evaluate the impact of these unknown quantities. The valuation exercise is an attempt to measure the total economic value related to deforestation in Amazonia, using the year 2000 as a benchmark.

Methodology and Estimates
The estimates were limited to particular elements for which available ecological information permitted conclusions to be drawn regarding the size and monetary value of environmental damage.

---

29. This section summarizes the results of an exercise contracted for this research with the same title—Seroa da Motta (2002). The complete study can be consulted on the World Bank Website (www.bancomundial.org.br).

No estimates were made of indirect uses such as maintenance of the local climate, erosion control, flooding, water recycling, and anti-fire protection.[30] The following elements ere estimated:

- Use values associated with timber extraction, non-timber extractivism and ecotourism;
- Indirect use values linked to carbon stocking;
- Option values associated with bioprospection;
- Existence value associated with biodiversity conservation.

It should be emphasized that the exercise was approached in terms of average rather than marginal values (as should be the case). Given the lack of information, this was the best result under the circumstances. In the estimates of foregone output, it was decided to adopt conservative average values or estimates referring only to certain areas but divided by the total area of Legal Amazonia (ecotourism). In the case of the option and existence values, parameters from the literature were employed which possibly dealt with marginal valuation.

Timber extraction
The only study available that analyzes the cash flow of forest management—the exploitation method on which other forest environmental services are preserved—is that which draws on experiments in the Paragominas region (Almeida and Uhl 1995). These authors concluded that it should be possible to generate a net income of US$28 per ha/year with these operations. Because this value related to 1994, it was updated to 2000, increasing to US$28.5 per ha/year. It is fairly plausible that this is an underestimate as far as forecasts of future values are concerned, since markets for Southeast Asian timber appear to be gradually drying up, thereby possibly resulting in higher prices for Amazon timber, which could include bringing onto the commercial market timber species which to date have not been marketed to any great extent. The value is well within the range presented in the literature review by Schneider et al. (2000).

Non-timber extractivism
Assuming that extraction of non-timber products is already practiced on a sustainable basis in the region, deforestation for cattle ranching will cause a production loss which, divided by the total area, will give admittedly very small values when compared with associated activities. Data referring to the municipal value of extractive production was used, based upon the municipal plant extraction figures of IBGE. Net income was assumed equal to gross revenue, considering that these are activities of low capital-intensity. Dividing the aggregate of the region by its total area, converted at the average exchange rate for the year 2000, resulted in a value of only US$0.20 per ha/year.[31] Unlike timber extraction, the future market for non-timber extraction activities is much less promising, especially when such activities over the whole of the region's territory are taken into account. In these circumstances, a reasonable assumption would be that it is unlikely that the benefits per hectare will increase substantially in the future.

Ecotourism
Ecotourism in Amazonia is an incipient activity which has not yet been systematically researched, and therefore little relevant data exist. Given the current size of the preserved area of Amazonia, ecotourism nevertheless has potential for growth even if the forested area were to suffer further reduction.

---

30. Fire used for opening up ranching areas is not only an agent of deforestation but leads to fire continuing under the forest litter (ground fire) thereby increasing the risk of devastating huge areas. The intensive use of fire for clearing pastures can accidentally cause damage to installations. The latter damage is regarded as the result of ranching activity and not of deforestation. For an estimate of accidental damage caused by fire, see Seroa da Mota et al (2001).

31. The value is high compared to the one presented in Wunder (2001) of US$ 0.7/ha/year practiced in 2-3 million hectares of the Amazon.

Estimates considered that the potential for ecotourism in Amazonia would be at most equivalent to the current potential of the same activity in the Pantanal region of Brazil, where a consolidated ecotourism sector exists in a biome which is almost totally preserved, is home to a wealth of biodiversity and water resources, and thus has great attraction for this type of tourism. It was also assumed that the growing demand for ecotourism could lead to growth in both biomes without one region affecting the other negatively. Using municipal data on maximum potential for ecotourism activity in the State of Mato Grosso do Sul, a figure of US$9 per ha/year was arrived at, representing the net loss of revenue that the non-development of ecotourism would incur in Amazonia.

Carbon stocking
It is difficult to quantify the forest carbon stock, above all in the Amazon forest, where a wide variety of different geographical features exist. Estimates for density cover a range between 70 and 120 tons of carbon per hectare (Rovere 2000). A density of 100 tons of carbon per hectare has been adopted here—lower than the figure of 191 tons per hectare used by Fearnside (1997) and than the 150 tons per hectare used by Andersen et al. (2001). This probably better represents the average density of the region, considering that in the transitional areas (with less biomass) deforestation is more pronounced.

There is also considerable argument about the actual value of a ton of carbon. The smaller estimate of $3 per ton of carbon suggested by UNCTAD (2001) is used here. The latter is derived from the most recent models which estimate the equilibrium price of carbon based on the Kyoto Protocol instruments. The cost control curves for each country show a price range of between US$3 and US$10 per ton.

Taking into account the average carbon density of 100 tons of carbon per hectare and a price of US$3 per ton, the value of carbon would amount to US$300 per hectare. Since this is a present value insofar as the opportunity cost of the carbon would be payment for foregoing it in perpetuity, a discount rate of 6 percent was used to annualize it, giving an annual value of US$18 per hectare—certainly a conservative estimate.[32]

Bioprospection
The prospect of the forest biodiversity yielding new drugs and their active principles for medical uses has been considered as one of the main incentives for preserving the Amazon forest. However, it is difficult to estimate the value of this potential benefit because it would require knowledge not only of the biodiversity itself but also of the economics of *bioprospection*. The few available studies that have been carried out differ radically in their estimates—anywhere between US$0.01 and US$21 per ha/year (Pearce 1993). It is not intended to present here the different methodologies used, nor the results. This study adopts the higher value put on it by Pearce (1993) of US$23 per ha/year, as the consumer surplus in this case seems to be closer to gains in terms of well-being experienced by the population groups benefited. These values have not been adjusted for the year 2000.

Existence value
Estimates related to the "existence value" associated with preservation (non-use) of tropical forests show a wide variety of values in the literature. The studies carried out to date tend to be based upon contingent valuation in rich countries where people appear to be willing to pay for

---

32. This estimate is much lower than those adopted by Andersen et al (2001) of US$45.00 per hectare per year and Fearnside (1997) of US$70.00 per hectare per year, since both authors adopt higher values for density and for the opportunity cost per ton of carbon—values which were accepted at the time that these studies were carried out. The estimates presented are closer to the amount of net income foregone on carbon sales and therefore capable of being absorbed in the local economy in the event of a market emerging.

the costs of preserving natural species and places. Horton et al. (2002) have produced the most recent study on the topic. The contingent valuation is applied to the specific case of the willingness to maintain conservation units in Amazonia detected among a sample of people in the United Kingdom and Italy. Two possible conservation scenarios are presented, based on conservation values of 5 percent and 20 percent. The study identifies an annual value in the form of an additional tax in each country and not a single fixed value to be allocated by an international fund. The average value estimated, combining the samples in both countries, was US$50 per ha/year for 5 percent of the area of Amazonia and US$67 per ha/year for 20 percent conservation. When the order of the questions was inverted (first 20 percent, followed by 5 percent) the average estimates changed to US$36 per ha/year and US$50 per ha/year. This study used the latter, lower values.[33]

The reference suggests however that the estimated value embraces both indirect and "existence" values. Moreover, only part of the forest area is valued, meaning that the measured value cannot be applied to the deforestation of one hectare. A number of adjustments were therefore made to these values: (i) estimating their equivalents for the rest of the world population; (ii) isolating the "non-use" value; (iii) extrapolating it to the total forest stock and (iv) aggregating it to apply to the total world population. The final result provides a value of US$31.2 per ha/year.

Summary of Estimates and Conclusions

Table 22 summarizes the estimates of the economic cost of deforestation in Amazonia both in values per hectare per year and as regards their respective present values when discounting at 10 percent, 6 percent and 2 percent per annum in perpetuity. The estimate of the total value is US$108 per ha/year. If this value were accepted by the region's landowners, it could make sustainable use of the greater part of the Amazon region viable. The value is slightly higher than that found in the main references on the basis of the same kind of evaluation (Fearnside 1997, Torras 2000, Andersen et al. 2002).

TABLE 22: SUMMARY OF THE ESTIMATES OF THE COSTS OF DEFORESTATION

Discount rate	Annual Value US$/ha/yr	10% p.a.	Present Value in US$/ha 6% p.a.	2% p.a.
Direct use value	37.7	377	628	1,884
Timber products	28.5	285	475	1,425
Non-timber products	0.2	2	3	9
Ecotourism	9.0	90	150	450
Indirect use value	18.0	180	300	900
Carbon stocking	18.0	180	300	900
Option value	21.0	210	350	1,050
Bioprospection	21.0	210	350	1,050
Existence value	31.2	312	520	1,560
Total (~)	108	1,080	1,800	5,4005

---

33. These differences on account of "ordering effects" cannot be analyzed by observing averages only but also variance components. In any event, to obtain an estimate of existence value using a decreasing scope valuation is more conservative.

TABLE 23: VALUE OF RENTING THE LAND FOR GRAZING IN LEGAL AMAZONIA, 1998-2001 (US$ PER HA/YEAR)[36]

	Rondônia	Acre	Amazonas	Pará	Tocantins	Maranhão	Mato Grosso
June 1998	43.2	—	77.6	84.2	30.6	44.3	49.0
June 1999	31.1	50.8	50.6	37.6	28.3	33.8	30.8
June/ 2000	40.1	49.5	—	41.0	33.4	34.0	32.6

Note: Values converted at commercial exchange rate of the month.
Source: Fundação Getulio Vargas.

Sustainable Alternatives: Comparing Costs and Benefits

This final section compares results of this chapter with those of Chapter 4. Some comparisons are both feasible and relevant regardless of the fact that this study cannot be termed a full cost-benefit analysis, in view of the caveats regarding the values calculated and the fact that the majority of the estimates were in terms of average rather than marginal values. However, the comparisons provide some indication of orders of magnitude involved, as well as background information on a number of possible policy recommendations. It is also interesting to compare the marginal net costs and benefits of cattle ranching with some of the other activities which may be more competitive from the social point of view, in effect, when the social costs of both activities are taken into account. While this study has not analyzed alternative activities, it is possible to make comparisons with the results presented in other research, at least in the specific case of sustainable forest management.[34]

Renting

First, taking into account the estimates of the economic cost of deforestation, it can be observed that the direct use values of the standing forest (obtained directly by the local population) amount to US$37.7 per ha/year, or 35 percent of the total value. Of these, only US$28.7 per ha/year arise from extractive activities. A first comparison is between these costs and those involved in renting the land for pasture.[35] Rent represents the income that the local producer would forego by not turning the land over for clearing and cattle ranching.

As can be seen in Table 23, the rental values for the year 2000 varied from US$32.6 to US$49.5 per hectare per year—above those estimated for direct use of between US$28.7 and US$37.7. Considering the uncertainties and the transaction costs involved in adopting new forest exploitation practices, the returns from the latter would not be enough to provide an effective incentive for local producers

---

34. Since a full benefit-cost analysis has not been achieved, it might be desirable to pre-establish levels of "tolerable" or acceptable deforestation and to estimate the costs involved—a cost-effectiveness analysis—an exercise that would not be significantly simpler than the estimates made here.

35. It can be observed that it is not appropriate here to draw comparisons of land prices in the region with the estimates of present value of the economic cost of deforestation, since these land prices, in addition to being of uncertain value in terms of real stock and share prices in the country's overall economy, reflect a private discount rate which is conceptually different from the social rates adopted in the estimates presented here. However, so long as deforesting can be financed from the sale of felled timber, the deforested land could be used for pasture.

36. Note that values in US$ in the table decreased from 1998 to 2000. This reduction was due to one single cause—devaluation of the Brazilian currency in January 1999. The values in Reals increased in the majority of states during the period analyzed in the table. Nevertheless, since the exchange rate in year 2000 was close to the real exchange rate, the values for 2000 do not include exchange rate compensation.

to adopt them. This result has to be regarded with caution: given the uncertainty of the various estimates, the differences between the rent and direct use values are likely not significant.

The previous comparison is restricted to direct use values. If the indirect use value of carbon stocking were also taken into account, it would add a further US$18 per ha/year to the income of the local producer, making sustainable use of the forest viable. However, payments for indirect use have yet to be incorporated into the markets to benefit local producers. The issue is one of economic policy concerned with the establishment of these mechanisms (and their transaction costs), not of the actual values involved.

The cost estimates involving both direct and indirect uses take into account relative prices and the current forest stock. As pointed out in previous sections, it can be reasonably surmised that the value of timber, as well as that of ecotourism and carbon, will appreciate over time as the sources of such goods and services get scarcer. If development policies for the area were to take into account technical training, favorable relative prices and expansion of the market for environmental services, the appropriation of these values by local producers (as the result of national and international initiatives) could generate an additional net annual return of almost US$56/ha of foregone deforestation and at the same time make sustainable productive activities in the area entirely viable.

Incentives for sustainable forest use would be even greater if option and existence values were also incorporated. If the estimated values (US$21.0 and US$31.2 per ha/year respectively) were realistic and if international compensation instruments aimed solely at conserving the Amazon forest were created, significant monetary incentives for controlling the process of deforestation would be generated. As in the case of indirect costs, the difficulty again revolves around the practical issue of creating workable financial transfer mechanisms.

In short, the estimates presented appear to confirm that important trade-offs exist between current and sustainable use of forest land. However, to avoid that losses are totally borne by the local community and to focus on such losses as a stimulus for introducing changed land use patterns, new market mechanisms or schemes involving international compensation need to be created, adding value to the environmental services of the forest that can benefit the world population as a whole.

Cattle Ranching

Regarding the results of cattle ranching as practiced in the area, Chapter 4 presented a model of linear programming focused on risk minimization, based on fieldwork estimates of the net profits to be obtained from soybean, corn, rice and beef production. On a representative establishment in the region (15,000 hectares with 30 percent cleared land), the model demonstrates that if the risk of cattle ranching were stochastically incorporated, a payment of at least R$45 per ha/year would make it feasible for the farmer to retain the forest area regardless of all other possible agricultural activities. In other words, none of these activities would generate an expected profit of more than R$45 per ha/year. If risk is ignored, the opportunity cost of forest conservation for ranchers increases to as much as R$200 per ha/year, although in this extreme case land would be allocated entirely to soybeans and forests, and not to pasture.

These values need to be compared with the economic costs (excluding direct costs) in order to evaluate required compensation from an external source and not from income arising from alternative uses (already analyzed). In this event, the values were estimated to be R$128 per ha/year,[37] representing forest services for global benefit made available as the result of international compensation mechanisms.[38] These are average, deterministic values and should be compared

---

37. The R$ 128 correspond to the US$ 70.2 presented in the previous section converted to Reals at an exchange rate of US$ 1 = R$ 1.83 in the year 2000. Such conversion should consider an exchange rate risk, although the estimates are made for the year 2000, when a slightly overvalued Real prevailed. Of the R$ 128, R$ 33 are from carbon, R$ 38 from bioprospection and R$ 57 for "existence" of the forest.

38. Payments for private conservation services to avoid deforestation were not particularly successful in an experiment in Costa Rica, when the value of such compensations did not keep up with the opportunity costs of cattle ranching (see Segura-Bonilla 2000).

therefore with the opportunity cost of the forest in the case that ignores risks (R$200 per ha/year). But as just indicated, in this extreme case farmers would convert the entire area to soybeans and not to cattle ranching, making comparisons difficult. *The result suggests that the sum of the economic costs of deforestation may or may not cover the opportunity costs of cattle ranchers, depending on their degree of risk aversion.* In practice, such degree of risk depends on the location of the property and the type of activity being considered. In the more consolidated areas, with greater presence of agriculture, the result would not be encouraging from the point of view of environmental conservation. An increase of about 50 percent of the values presented here would be called for, or the inclusion of other environmental services not valued here (indirect uses) in order to provide enough social benefits to produce sufficient social (global) benefits to compensate for the farmer's losses (note again that this is a farmer maximizing profits and choosing between forests, crops and cattle). In the areas where cattle predominates, the potential for compensation based on the indirect use values is more than sufficient to cover the opportunity costs of producers. The figures need to be taken as orders of magnitude. More careful examination suggests that the values put on ranching and deforestation are relatively close. Further analysis, especially regarding environmental costs, is called for in order to arrive at a firmer conclusion.

Forest Management

Rather than pure conservation, landowners might decide to opt for extracting products from the standing forest. One possible sustainable activity in Amazonia is timber exploitation, based on rotation of cutting areas and low impact management practices. Barreto (2002) estimated that the net present value of sustainable timber extraction for 1998 was US$203/ha. This can be compared with the net income from cattle ranching identified in Chapter 4—in the region of R$100 per ha/year, measured in 2002. Using the same discount rate of 8 percent as that applied to timber and using an approximate exchange rate of R$2.50 to the US$, net present value for cattle ranching is US$500/ha. This is almost identical to that presented by Barreto for ranching—more than double that to be secured from forest management. It must be emphasized that Barreto claims that the current net values of unmanaged forest activity are practically the same as for managed activity. Seen from any angle, *forest management, unlike cattle ranching, is not a financially attractive proposition.*

This result reflects what is being observed on the ground—that cattle ranching in Amazonia is quite profitable, and that sustainable alternatives such as forest management are unable to compete *from a private producer's point of view*. In addition to the forest management not being competitive, both the risk factor and the institutional set-up cannot be understated.

From the social point of view, however, since forest management leaves the forest system largely unaffected, the social benefits form forest services remain essentially unaltered. Thus, as in the case of the comparison with ranching, to the net present value of US$203/ha should be added the value of indirect use, option and existence values (US$70.4 per ha/year) which, using an 8 percent discount rate produces a present value of US$875/ha. Added to the direct use value of US$203, an approximate value of US$1,100/ha emerges—double the value of cattle ranching. This indicates that *forest management yields larger benefits from the social point of view than cattle ranching, even though ranching generates higher private financial returns.* This result is crucial to the policy recommendations put forth in the concluding chapter.

An additional observation needs to be made with regard to the competition and divergence between the private and social costs and benefits of cattle ranching, sustainable forest management (SFM) and unsustainable logging. While there is little contention that SFM produces a socially more desirable outcome, this study suggests that from a private perspective it cannot compete with cattle ranching, while other studies suggest that it produces roughly the same rates of return as unsustainable logging. While it is common to have unsustainable logging followed by cattle ranching, SFM is an exclusive activity. Therefore, *SFM actually has to compete with the sum of the two activities, so that the gap between the private and social returns on forest land makes it almost impossible for the socially preferable activity to be implemented without some form of government intervention.*

Lastly, it should be mentioned that all the above analyses should be regarded considering the limitations imposed by the conditions under which they are made. The proposition that sustainable forest management is superior to cattle ranching does not mean that all forest outside of strict conservation units should be put into sustainable logging, because certain forest regions may have low returns on account of the lack of more noble species of trees or their distance from markets. On the other hand, the geo-ecological conditions of certain regions may be unsuitable for cattle ranching and appropriate for sustainable logging, so that even from the private perspective sustainable logging is more viable. The scale and type of ranching are other critical factors. Other production systems on a smaller scale may produce different results (Carpentier et al. 1999).

Chapter 6

Conclusions and Recommendations

This section summarizes the main results and conclusions of the present study and makes a number of policy suggestions.

Main Results: Conclusions

- The basic conclusion of the study is that deforestation in Amazonia is not a classic "lose-lose" situation characterized only by economic and environmental losses. The process involves a series of tradeoffs, with obvious private economic gains.
- Land-use data on Amazonia demonstrates that the main cause of deforestation in the region is cattle ranching. Expansion of ranching since the early 1970s has been a continuous and inertial process.
- Remote sensing data, together with IBGE figures, suggest that the large and mid-size agents are primarily responsible for deforestation in the region. The smaller agents are used as labor or for helping to consolidate land holdings and their possession (the so-called "warming" process). They tend to make only a minimum direct contribution to deforestation. Moreover, deforestation by small agents is more acceptable from a social point of view because it probably leads to improvements in living conditions for poorer population groups, which is not the case of deforestation by larger agents.
- In spite of their different motivations, interests and economic strategies, the groups of social actors on the frontier depend on each other. The profits obtained by speculators on the newer frontiers ultimately depend on the activities of the more capitalized and professional ranchers on the consolidated frontier who are determined to continue expanding their businesses.
- Regardless of who the original agents are, the end result of the land occupation process is almost inevitably the establishment of ranching activities. If this were not financially viable,

forest conversion or deforestation would not occur on the scale on which it is occurring since the initial agents would hardly be able to recover the costs involved in deforesting and preparing the land.
- On the speculative frontier the State has only a reduced presence. Such public authorities as exist are beholden to local elites with speculative interests—aimed primarily at encouraging the opening up of new areas in order to expand their ranching activities.
- The small local actors in these areas are not particularly interested in protecting the forest. Rather, their main concern is to ensure possession of their land, to protect themselves from rural violence and to confront the monopsonistic conditions that they face when selling their product.
- The economy of the agents on the speculative frontier is based more on the sale of land than on the returns to be obtained from ranching. In such areas, the main purpose of cattle ranching for such people is simply to ensure property rights. As indicated in the field work in the advanced frontier, the rate of return on such activities is below five percent.
- The combination of the potentially high private profitability of cattle ranching with affordable transport costs (existence of roads) leads to deforestation. So long as geo-ecological conditions are favorable, there will always be pressure for road building (endogeneity) to the point that ranchers will often themselves build the road network. If cattle ranching were not privately profitable, the existence of roads *per se*, or the roads built with more geo-political objectives (the "exogenous roads") would not lead to so much deforestation and land conversion. Undoubtedly, however, penetration roads built into areas with little occupation clearly lead to increased deforestation.
- Conditions for raising livestock in Amazonia are surprisingly favorable, mainly in the already occupied regions, largely as a consequence of the precipitation levels, temperature, air humidity, and types of pasture.
- The rates of return on ranching itself (excluding sales of timber for example) calculated at different points on the arc of deforestation, are consistently above ten percent—much higher than those found in the rest of the country. These are not average values for the region but can certainly be achieved by the more professional and better capitalized ranchers.
- The level of professionalism of producers on the consolidated frontier leaves no doubt as to these agents' concern with the constant need to modernize, expand markets, improve production expertise, and to employ pasture and animal management techniques—in other words, to stay competitive through greater efficiency.
- There are few doubts as to the sustainability of production (understood only in the sense of the feasibility of maintaining the activity over a long period). However, it would be premature to be overly optimistic, given the limited time that cattle ranching has been practiced in the region. Our study pointed to a number of positive as well as negative factors that could affect productivity over the longer term. Despite the uncertainties it seems to be a fair assumption that cattle production is increasingly sustainable in the region.
- In the short term, there appears to be no trend pointing towards a decline of the expansion of the cattle ranching frontier into native forest—at least in the areas with production conditions similar to those analyzed in the study. Even without government subsidies, the profitability of the ranching sector is the driving force behind the inertial process.
- Agriculture cannot compete with cattle ranching in the forest areas. Geo-ecological barriers are in general more restrictive in the case of agriculture. A prime example is the high precipitation in certain areas (above 2000 mm/year). In the areas with less than 2000 mm/year rainfall, it is one of the factors which most favors ranching, which is precisely why it predominates in such areas.
- In the course of examining representative establishments on the consolidated frontier, the study obtained economic and financial information and developed simulation models of ranching activities, demonstrating that the producers are averse to risk and tend to avoid

specialization, preferring to work with a combination of crops, pasture and forest where they are each possible.
- The simulations show that established producers would be prepared to accept relatively low sums (R$45 per ha/year) as compensation for foregoing expansion into the forest of cultivated areas. These amounts could be as high as R$200 per ha/year in cases where there is no risk aversion (in which case producers would convert forest to agriculture, and not to cattle ranching, under the model's assumptions).
- The simulations made of charging taxes on deforestation suggest that the imposition of a US$15-20 per ha/year tax would not reduce significantly the deforested areas: producers would tend first to change the mix of crops, as opposed to reducing the amount of forest clearance. The two policies are equivalent, and differ with respect to who bears the costs.
- Government subsidies and credits for cattle ranching tailed off significantly in the 1990s, but this had virtually no effect on deforestation. Currently, such incentives cannot be considered any more as relevant factors which explain the deforestation process in Amazonia. The preferential FNO remains a social program which, despite the difficulties, has led to improvements for certain small producer sectors.
- Since 1970, regional income has risen substantially. Rural income per capita in particular tripled on average from US$410 in 1970 to US$1,417 in 1995. In the states with the highest deforestation rates, the increase was even higher.
- Increased per capita rural income does not necessarily translate into quality of life improvements for the poorest local population groups. The social gains observed during the past three decades in the region are difficult to interpret from available data. Socio-economic indicators show there has indeed been significant progress but it has been insufficient to reduce the gap in relation to the rest of the country (see annex).
- Moreover, the largest slice of regional income originated in the urban as opposed to rural sectors, which suggests that improvements in social conditions probably had little direct link with deforestation. When calculating social benefits, therefore, improvements in Amazonia cannot be credited predominantly to deforestation. Private benefits from large scale cattle ranching are mostly exclusive, probably having contributed little to alleviate social and economic inequalities. The available data does not allow for firmer conclusions—there is indeed a good case for conducting additional research on this specific topic.
- The study notes, however, that there have been social benefits that go beyond the sectoral and regional boundaries. At the national level, cattle ranching in Amazonia has allowed beef prices to consistently fall in the last 5 years, when 100 percent of the growth of the national herd took place in the states of Pará, Mato Grosso, and Rondônia—the three champions of deforestation in the Amazonia. At the same time, beef exports jumped from 350,000 tons in 1999 to 900,000 in 2002, representing nearly US$1 billion in foreign exchange earnings.
- The social costs of deforestation were estimated to be around US$100 per ha/year. This value is open to doubts, arising from the constraints imposed by the methodologies used in environmental valuation and the limitations of the data available. However, the value exceeds the potential income to be derived from cattle ranching per hectare and could form a basis for compensation purposes. Since transfer mechanisms do not exist which would make such compensation feasible, the main point at issue is the private profit derived from cattle ranching—which is entirely positive and which, as the study suggests, constitutes the key factor explaining the deforestation process in Brazilian Amazonia.
- In comparison with sustainable forest management (SFM), ranching is more economically viable *from the private point of view, although we must be cautious with a definite conclusion*. It is also a low-risk venture according to the overwhelming evidence from farm interviews. SFM on the other hand is a poorly disseminated and rather "sophisticated" technique that has to compete with both unsustainable logging as well as cattle ranching (since cattle

ranching can and typically does succeed unsustainable logging, the total return on the land is the sum of the returns of the two activities). SFM excludes other land uses, which makes it even more difficult to compete with the traditional activities. This may largely explain the ongoing deforestation and the expansion of ranching activities in the region.
- However, *from the social point of view* the study suggests that a higher economic benefit can be obtained from SFM. Sustainable forest management can also be assumed to be better from an environmental and social standpoint (despite some skepticism with regard to the later by Wunder 2001). An institutional set-up to change such scenario is currently missing.
- Finally, the various analyses carried out in the course of the study need to be viewed in context—mainly as regards the private economic viability of ranching which, in principle, is valid only in the areas studied and under specific conditions. The key conditions are the level of professionalism of the ranchers, the amount of rainfall and most importantly the scale of production (size of holdings). This last factor is critical for the viability, and thus for the expansion of cattle ranching in Amazonia. At the local level, a number of results could diverge from those presented in the present study.

Recommendations

The following recommendations are based on the results of the study. The main objective is to support sustainable development policies for the region. Unless otherwise noted, the recommendations apply equally to the Brazilian Government and the World Bank, in view of the congruity of their objectives.

Information and Planning

Drawing up and agreeing on sustainable development strategies requires the identification of the principal social agents, and of their manifold and conflicting interests and motivations. This study suggests that a fundamental step in this direction would be to accept the thesis that *cattle ranching in Amazonia is a potentially profitable activity for producers and that profitability is the basic driving force behind the deforestation process in the region.* This implies, furthermore, acknowledging that tradeoffs exist in the process of deforestation in Brazilian Amazonia.

Forest protection policies should as a priority perhaps be aimed at producers on the consolidated frontier who are the driving force of the process and not merely at those on the speculative frontier. Obviously, this does not mean that there should be no effort to enforce the law and control the illegal operations of the agents on the speculative frontier. But the weak presence of government authorities, together with the low degree of risk aversion of these agents, suggests that the focus of policies should be concentrated upon the more professional agents operating on the consolidated frontier.

The strategy should be to work with cattle ranchers and not against them. While a large proportion of these agents may not be prepared or willing to negotiate, there are more amenable leaders interested in some sort of compromise with the government and society in order to have their activities legalized. Because many of those have close links with municipal governments in the region, the latter could perhaps represent them or at least participate in the negotiating process. This might be a good opportunity for the new federal administration (and possibly also for the states).

The authorities responsible for protection of the Amazon forest should increase their focus on ranchers as agents of deforestation. Currently, little attention is paid to cattle ranchers in comparison with loggers. Control over the loggers is fundamental not in terms of deforestation per se but because sustainable forest management represents the principal alternative activity to ranching in Amazonia capable of competing on the same scale. Disorderly, predatory and predominantly illegal timber extraction as currently practiced should be fought not only on account of its illegality but because it excludes the possibility of present or future implantation of an activity which would be better than cattle ranching as being sustainable from an economic, social and environmental point of view.

Zoning could be used as a negotiating process between the economic agents (including cattle ranchers) and the government, leading gradually to compromised land occupation in regions and areas which are suitable from a social, economic and environmental point of view. The World Bank has for a number of years supported zoning initiatives throughout the region and should continue to concentrate on their practical application, continuous updating and revision, and in particular on the process of negotiation between the participating actors. It should be emphasized that the small agents who are identified here as being of lesser importance to explain deforestation in Amazonia, should not be sidelined from this process. On the contrary, *because they are the main target for sustainable development, they should also participate in the process of determining regional development strategies.*

Since a great deal of ignorance and uncertainty exists about the various factors and effects associated with the deforestation process and the expansion of the frontier, *the risks involved suggest the adoption of conservative strategies*. The heritage at risk in Amazonia should not be endangered by irreversible decisions involving high social, economic and environmental costs. Thus, conservation initiatives being implemented should be encouraged. The World Bank continues to support such initiatives through the ARPA, PROBIO and PROARCO Projects, and various projects supported by the Pilot Program (Redwood 2002).

Among the most important factors analyzed in this study and on which it will be important to *increase our knowledge are the environmental values and services* of the forest (the social costs of deforestation), both with regard to technical information on the complex ecological effects as with regard to environmental valuation. This should include studies of the *possible social benefits linked with them* (that is, up to what point can improvements in socio-economic conditions in the region over the past few decades be directly or indirectly attributed to deforestation).

It is perhaps of equal importance to *analyze more thoroughly the effects of transport costs on deforestation*. This is important for evaluating both the possible effects of the *Avança Brasil* Program on deforestation, as well as the effects of smaller feeder roads, particularly in the more consolidated frontier. Such roads may enhance intensification but their final impact in terms of deforestation is less clear. The study also suggests that local roads should be understood more as endogenous to the deforestation process and not as something decided exogenously to local agents.

Economic Instruments

Creation of tradable development rights. A more flexible handling of the legal reserve requirements in private properties could yield enormous ecological and economic rewards. There is no reason why certain areas—typically the most fertile and productive areas—should not be allowed to benefit from higher percentages of deforestation provided that this is compensated with additional legal reserves in the ecologically richer areas. These areas could be indicated by zoning, and in principle there should be sufficient flexibility for tradable rights to be applied in other ecosystems and water basins, provided that certain rules of ecological endemism and other explicit and binding technical parameters are respected. Strict compliance with current law will be very difficult, involve enormous costs and yield doubtful environmental benefits: it would be unrealistic to expect that the properties in the *cerrado* of southern Mato Grosso, for example, will totally fulfill their legal reserve obligations. As an alternative, it may be of interest to require local producers (possibly acting in consortia) to compensate for reduced legal reserves with even larger areas in other ecosystems on the cerrado. The pilot experiment in Paraná is an example that might also be applied in Amazonia. It would be a question of perfecting the mechanism in the context of the Forest Code.

One of the classic economic solutions to the problem is to tax deforestation so as to force agents to internalize the environmental costs. The taxation simulations made for this study suggest that high taxes are necessary to reduce forest clearing significantly, but further studies need to be done to better evaluate the instrument under conditions different from those analyzed here. Taxes could be introduced in the context of the proposal to create an environmental contribution (*contribuição ambiental*) included in the latest tax reform proposal submitted by the Federal Government to the National Congress.

A theoretically equivalent alternative to taxation is to compensate agents for not deforesting (although the political implications of who pays are very different). The simulations suggest that established cattle ranchers would be prepared to accept relatively small sums for foregoing new deforestation, depending on their level of risk aversion. From a national point of view alone, legal precedents exist in Brazilian legislation where royalties are paid in the cases of oil extraction and the flooding of areas for the construction of hydro power dams. The values paid are not calculated on the basis of a rigorous monetary evaluation of the environmental costs, and there is no reason for not introducing the same as a form of compensation for the loss of biodiversity and of the innumerable forest environmental services.

In addition to the national interest, the international community also benefits from the environmental services of the forest. The study showed that the sum of national and international benefits may be higher than the returns from cattle ranching. However, *transfer mechanisms do not yet exist in practice* and serious technical and political difficulties need to be confronted in order for them to be implemented. The few experiences in other countries are not particularly encouraging. In spite of the difficulties, the World Bank probably has a role to play in assisting the Brazilian government to identify international initiatives and partners which could help to design the transfer mechanisms just mentioned. This would also be an appropriate role for the international donor community, given its concern with deforestation in the region.

The search for sustainable alternatives has been limited to forest management and small scale one-off initiatives, some of them socially, economically and environmentally superior to cattle ranching. These efforts should continue, keeping in mind, however, that while superior, they have not been able to compete in terms of scale with cattle ranching and have mostly remained at the trial level. The available scientific and technological resources applied to the region have been insufficient for obtaining greater knowledge and disseminating existing experiences. The Federal Government and the World Bank, through the Rain Forest Pilot Program, could play a greater role as catalyst in the dissemination of information.

The World Bank should review its conservative approach adopted in relation to Amazonia over the past decade and focus more on the promotion of sustainable development. This does not mean abandoning support to conservation but rather making the two approaches complementary. Promoting productive activities with high social and economic benefits could, with low environmental impacts (or even none at all), represent alternatives of major interest for development of the region. Approval of a project in support of the National Forest Program (under discussion with the Federal Government) would be an excellent step in this direction.

The fiscal incentives which benefited larger landholders have been reduced and now tend to be better applied. *Social programs such as the preferential FNO and the INCRA settlement projects could bring higher ecological and social gains, particularly for poorer and less well qualified landowners.* The marriage of interests between environmental protection and support for traditional local populations is one of the top socio-environmental concerns of the new federal administration and should be given every support.

Other economic instruments which have been the subject of discussion for some time among MMA (Environment Ministry), IPEA and World Bank officials include (see Seroa da Motta et al. 2000, Haddad and Rezende 2001): (i) the introduction of the ecological value-added tax (ICMS) following its successful implementation in some states (it is in the course of being implanted in Rondônia and Mato Grosso)—protection of areas of native forest could be one of the criteria for compensating municipalities in the calculation of the municipal value added; (ii) the introduction of environmental criteria similar to those of the ecological ICMS in the States and Municipalities Participation Fund (*Fundo de Participação dos Estados e Municípios*—FPE and FPM); (iii) reorientation of the criteria governing award of fiscal or credit subsidies to promote sustainable activities, development of sustainable technologies and scientific research; (iv) introduction of environmental criteria in the concession of agricultural credit in the region; (v) reviewing and eliminating existing subsidized credits for traditional cattle ranching in Amazonia.

Enforcing the Law
Whatever the economic incentives, they will certainly call for greater enforcement capacity. This is bound to be a major uphill struggle owing to the immense size of the region and the difficulties of working with the local agents. No matter how much political determination exists, it will always be difficult to stem the inertial trend observed over several decades. The balance between factors that in the future could favor even more cattle ranching in Amazonia over those that could impede it probably favors the former. However, the combination of market mechanisms and command and control is probably still insufficient to slow down or halt, in the short term, a process which has been observed over so many years. This further reinforces the need for negotiated solutions which take the cattle ranchers into account.

To ensure more effective action an institutional cooperation strategy is fundamental. Institutions such as the MMA, IBAMA, ADA, the Ministry for Regional Development (*Integração Nacional*), the Ministry of Planning, INCRA, FUNAI, and state governments need to work together, agreeing on common targets and defining individual functions. A clear and transparent identification of objectives and responsibilities is vital for each institution to have the appropriate incentives and to be accountable for its own performance.

Finally, in spite of the political difficulties, the process of conceding property rights needs to be urgently and seriously reviewed and audited. It is difficult to identify without a closer study the network of interests involved. The results—frequently associated with violence and fraud—are well known, but could be reverted if the agencies dealing with land occupation and property rights were to perform more effectively, bringing order to land use once and for all, protecting and lending support to small producers and guaranteeing the integrity of public land and of the natural and social heritage of the region. The speculative gains are colossal, and the key stage in the process is the regularization of property rights. There is no reason why the Federal Government, working in partnership with the states, could not take energetic action on this issue.

ANNEX

SOCIOECONOMIC DEVELOPMENT IN BRAZILIAN AMAZONIA

This Annex analyzes the evolution of selected socioeconomic indicators of Brazilian Amazonia in an effort to assess possible social gains associated with the process of deforestation and land occupation and the overall economic development of the region. Most of these changes result from a combination of factors, some of which could be directly or indirectly linked to deforestation and subsequent agriculture and cattle ranching activities; others are not connected to them. Because it is too difficult to establish causalities more firmly, it was decided to present these indicators independently and not as potential benefits from deforestation.

The analyses were conducted at the municipal level based on IBGE census data. Analyses at the municipal level may be too aggregated to permit precise evaluation of the social welfare aspects of local populations and to make comparisons of the changes within the municipalities and between one municipality and another over time. Nonetheless, the municipal data provided by IBGE is useful basic information, essential for trying better to understand the evolution of the local population's social and economic conditions.

Figure A1 shows the evolution of the percentage distribution of the population of Legal Amazonia, according to municipal income (GDP) per capita for the census years between 1970 and 1995, comparing it with per capita income for the whole of Brazil during the same period. The figure demonstrates that there was a significant improvement in income levels of the population of Amazonia over the past three decades. In 1970, about 75 percent of the population of Amazonia had a per capita income of less than US$1000. This percentage gradually declined over subsequent years, reaching a low point of 30 percent in 1995. Compared with the whole country, in 1970, 100 percent of the population of Amazonia lived in municipalities with an income of less than the national average, whereas the percentage fell back to

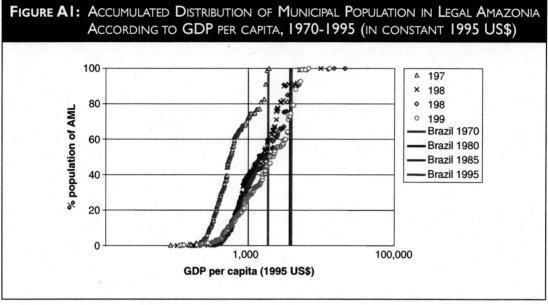

Source: IBGE, drawn by IPEA.

90 percent in 1980 and to 80 percent by 1995, representing a not insignificant improvement in income levels.[39]

From the temporal point of view, the figure shows a substantial improvement in per-capita income levels between 1970 and 1980, a period during which the Brazilian economy grew rapidly. From 1980 onwards, the economy stagnated and per-capita income improvements in Amazonia slowed down (and even receded slightly in some of the poorer municipalities). Nevertheless, over the same period, 80 percent of the region's population benefited significantly from rises in per-capita income, especially when compared with the almost imperceptible improvements at national level.

Figure A2 shows the incidence of poverty among the population of Amazonia. This suggests that poverty indices improved appreciably in the 1970s and 1980s. As with per capita income, the proportion of the population groups with insufficient income worsens during the 1980s. This could be partly due to the fact that 1980 represented a year of peak growth (due to negative real interest rates, a civil construction boom, etc.) and 1991 a year of stagnation caused by a decline in economic activity which followed sequestration of assets under the Collor Plan. Because more recent figures are not available, it is difficult to draw more definitive conclusions regarding poverty reduction in the region. The improvements observed in Legal Amazonia were very much a reflection of those observed throughout Brazil as a whole.

In addition to income, other socio-economic indicators were studied. Figures A3, A4 and A5 show the evolution of three classic indicators (life expectancy, infant mortality, and illiteracy rates). They follow the same format, and the interpretations are roughly similar. Figure A3 suggests, for example, that in 1970, 60 percent of the local population lived in municipalities where life

---

39. The figure implies that 15 municipalities in 1995 had a higher income than the national average. These in fact are not municipalities but Minimum Comparable Areas (MCA), as introduced in Chapter 3. Some of the municipalities have much greater weight than others in the same MCA. The 15 MCA with an income above the national average include 5 state capitals—Belém, Manaus, São Luis, Cuiabá and Rio Branco, 5 in the south of Mato Grosso and Tocantins (General Carneiro, Alto Graças, Itiquira, Alto Araguaia, São Miguel do Araguaia), in addition to Paragominas, Santa Isabel, Afuá and Almerim (Pará) and Bacarena (Maranhão).

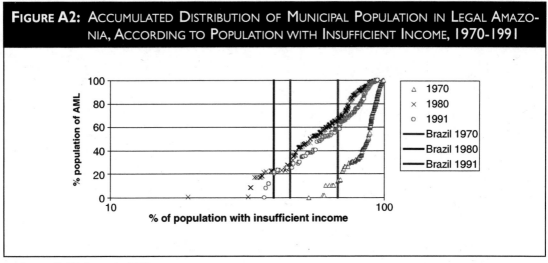

FIGURE A2: ACCUMULATED DISTRIBUTION OF MUNICIPAL POPULATION IN LEGAL AMAZONIA, ACCORDING TO POPULATION WITH INSUFFICIENT INCOME, 1970-1991

Source: IBGE, drawn by IPEA.

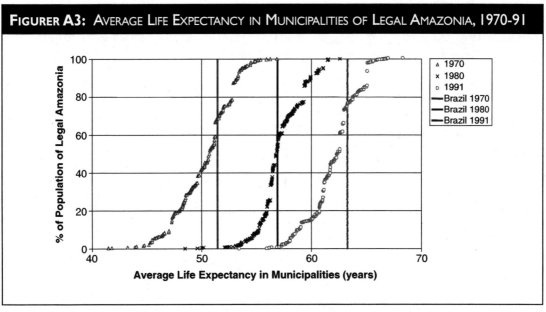

FIGURER A3: AVERAGE LIFE EXPECTANCY IN MUNICIPALITIES OF LEGAL AMAZONIA, 1970-91

Source: IBGE, drawn by IPEA.

expectancy was 52 years. In 1991, the same 60 percent of the population lived in municipalities with a life expectancy of 63 years. The figure also demonstrates that life expectancy increased in the region at a relatively slower rate than the national average: in 1970 and 1980, 60 percent of the municipalities of Amazonia experienced lower life expectancy rates than the national average; this percentage increased to around 75 percent by 1991.[40]

Infant mortality (Figure A4) declined significantly over the period under study. In 1970, fifty per cent of the local population inhabited municipalities with 120 infant deaths per 1,000 live

---

40. One possible explanation for this trend is the fact that the proportion of the rural population in Legal Amazonia exceeds the national average and is moreover highly vulnerable to tropical diseases.

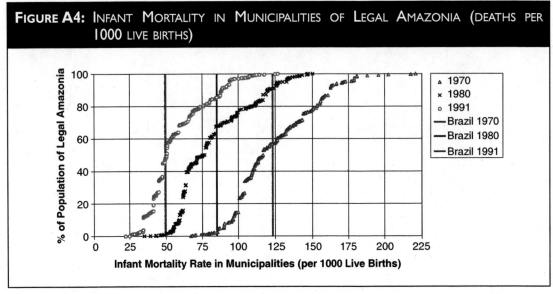

FIGURE A4: INFANT MORTALITY IN MUNICIPALITIES OF LEGAL AMAZONIA (DEATHS PER 1000 LIVE BIRTHS)

Source: IBGE, drawn by IPEA.

births. In 1991 the proportion fell to 50 deaths per 1000 live births. During the 1970s, the region experienced faster improvement in this respect than Brazil as a whole, although this was not the case in 1991. In 1970, 55 percent of the population lived in municipalities with infant mortality rates of less than the national average, whereas the percentage increased to 65 percent in 1980, and declined to just under 50 percent in 1991.

Finally, Figure A5 shows that the evolution of illiteracy rates practically equaled that of the country as a whole. Throughout all the periods observed, around 35 percent of the population lived in municipalities in Amazonia that had illiteracy rates of less than the national average. However, while in 1970 half the population lived in municipalities with illiteracy rates of over 45 percent, by 1991 the same proportion of the population lived in municipalities where the rate had fallen back to 25 percent.

The data in Figures A1-A5 suggest that social and economic conditions improved for the local population, although certain caveats apply. The graphs suggest furthermore that the region's improvements generally reflected the improvements registered in Brazil as a whole and were not significant enough to narrow the gap in relation to the rest of the country.

As regards income sources, the above data could in effect mask a possibly more substantial contribution made by the urban sectors. From 1985 onwards, the GDP of Legal Amazonia began to reflect a significantly larger contribution of income derived from urban as opposed to rural areas (see Table A1). The data suggest that the social improvements indicated by the previous graphs might not be linked to deforestation.

The Table suggests the possibility that recent socio-economic improvements in the region are derived not so much from cattle ranching but from the urban sectors. It is not possible, however, to state conclusively that this is the case on the basis of the data consulted. It could be said, with some justification, that many of the improvements were also due to the ranching sector.[41] Andersen et al. (2002) suggest that urbanization in the region was driven, in turn, by the expansion of

---

41. It is worth noting that despite the precise definition of IBGE, the concept of "urban" in Amazonia must be analysed with caution: since the minimum size of a city is the same as in the rest of the country, the "cities" in Amazonia have a much stronger link with rural areas and agriculture and cattle ranching activities than the Center-South of the country. Therefore, what is eventually considered as urban population or income may in fact be rural.

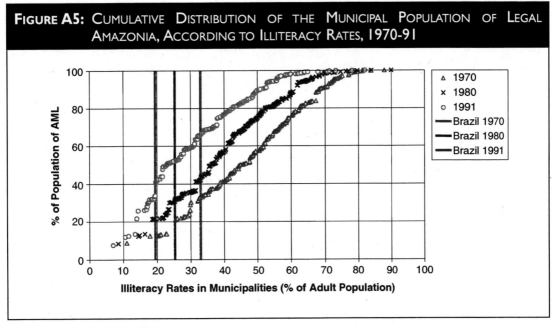

FIGURE A5: CUMULATIVE DISTRIBUTION OF THE MUNICIPAL POPULATION OF LEGAL AMAZONIA, ACCORDING TO ILLITERACY RATES, 1970-91

Source: IBGE, drawn by IPEA.

TABLE A1: EVOLUTION OF RURAL AND URBAN POPULATIONS, GDP AND GDP PER CAPITA IN LEGAL AMAZONIA, 1970-95

	Population (1000 inhab.)		GDP (1995 US$ million)		Per capita GDP (1995 US$)	
	Rural	Urban	Rural	Urban	Rural	Urban
1970	4,589	2,674	1,606	4,243	349	1,586
1975	5,263	3,631	2,460	7,535	467	2,075
1980	6,092	5,112	4,192	16,561	688	3,239
1985	6,638	6,531	5,215	23,710	785	3,630
1990	NA	NA	NA	NA	NA	NA
1995	7,000	10,692	9,882	33,948	1,411	3,175

Source: IBGE, drawn by IPEP.

the cattle ranching sector. In the case of income from the livestock sector, it would be interesting to make a distinction between income generated by larger ranchers from that earned by smaller operators—the former no doubt disposed of their earnings in a variety of ways, while in the case of the small ranchers it is a reasonable assumption that the bulk of their earnings was directed wholly towards improving their own living standards (Haan et al. 2001)—in other words, opportunities for "social mobility."

The available data also show that the socio-economic improvements were sufficient only to keep up with improvements recorded in the rest of the country. Changes in relative poverty levels in the Amazon compared with the south of the country point to a certain degree of progress, insofar as the inter-regional gap has not widened. However, there is no cause for celebration as the region continues to lag behind the rest of Brazil in terms of average socioeconomic conditions.

REFERENCES

Almeida, A.L.O. and J.S. Campari. 1993. *Sustainable Settlement in the Amazon.* Education and Social Policy Department, The World Bank, Washington D.C.

Almeida, O. T. and C. Uhl. 1995. "Identificando os custos de usos alternativos do solo para o planejamento municipal da Amazônia: o caso de Paragominas (PA)." In P. May, ed., *Economia Ecológica*. Rio de Janeiro: Ed. Campus.

Alves, D. 2000. "Distribuição Espacial do Desflorestamento na Amazônia Legal." Prepared for Secretaria de Coordenação da Amazônia do Ministério do Meio Ambiente, São José dos Campos.

Alves, D. 2001. "Space-Time Dynamics of Deforestation in Brazilian Amazônia." Draft, INPE, São Paulo.

Alves, D., W.M. Costa, M.I.S. Escada, E.S.S. Lopes, R.C.M. Souza, and J.D. Ortiz. 1997. "Análise das Taxas de Desflorestamento dos Municípios da Amazônia Legal nos períodos 1991-1992 e 1992-1994." Relatório Técnico AMZ-R02/97. Ministério da Ciência e Tecnologia, Ministério do Médio Ambiente, dos Recursos Hídricos e da Amazônia Legal. Brasília.

Andersen, L.E. 1997. "A Cost-Benefit Analysis of Deforestation in the Brazilian Amazon." IPEA, Texto Para Discussão No. 455, Rio de Janeiro.

Andersen, L.E. and E.J. Reis. 1997. "Deforestation, Development, and Government Policy in the Brazilian Amazon: an Econometric Analysis." Texto Para Discussão No.513, Rio de Janeiro.

Andersen, L.E., C.W.J. Granger, E.J. Reis, D. Weinhold, and S. Wunder. 2002. *The Dynamics of Deforestation and Economic Growth in the Brazilian Amazon.* London: Cambridge University Press.

Andrae, S. and K. Pingel. 2001. "Rain Forest Financial System: the Directed Credit Paradigm in the Brazilian Amazon and its Alternative." Mimeo, Institute of Latin American Studies, University of Berlin.

Angelsen, A. and D. Kaimowitz. 2001. *Agricultural Technologies and Tropical Deforestation.* Cabi Publishing (UK) and CIFOR (Indonesia).

Arima, E. 2000. *Incentivos Fiscais e de Crédito para Pecuária na Amazônia Legal.* Instituto do Homem e Meio Ambiente na Amazônia, Belém, Pará.

Arima, E. and C. Uhl. 1996. *Pecuária na Amazônia Oriental: Desempenho Atual e Perspectivas Futuras.* IMAZON, Série Amazônia No.1, Belém, Pará.

Arima, E. and C. Uhl. 1997. "Ranching in the Brazilian Amazon in a National Context: Economics, Policy, and Practice." *Society and Natural Resources* 10(5): 451-433.

Barros, G.S.C., S. Zen, M.R.P. Bacchi, S.M. Ichihara, M.Osaki, and L.A. Ponchio. 2002. *Economia da Pecuária de Corte na Região Norte do Brasil.* CEPEA/ESALQ-USP, Piracicaba, São Paulo.

Barreto, P. 2002. "Estudos sobre manejo florestal na Amazônia Brasileira." Relatório Técnico, Banco Mundial, Brasília.

Becker, B. 1999. "Cenários de Curto Prazo para o Desenvolvimento da Amazônia." Cadernos do NAPIAm, No.6, Brasília.

Binswanger, H.P. 1991. "Brazilian Policies that Encourage Deforestation in the Amazon." *World Development* 19(7): 821-829.

Browder, J.O. 1988. "Public Policy and Deforestation in the Brazilian Amazon." In R. Repetto and M. Gillis, eds., Public *Policy and the Misuse of Forest Resources.* Cambridge University Press.

Carpentier, C.L., J. Witcover, and S.A. Vosti. 1999. "Smallholder Deforestation and Land Use: A Baseline." Draft Technical Note, OED, World Bank, Washington, D.C.

Castro, E.R., R. Monteiro, and C. P. Castro. 2002. "Atores e Relações Sociais em Novas Fronteiras na Amazônia: Novo Progresso, Castelo de Sonhos e São Félix do Xingú." Background paper to the present report. Belém.

Castro, N.R. 2002. "Um Programa de Cálculo dos Custos de Transporte no Brasil." Database unpublished obtained directly from the author.

Cattaneo, A. 2001. "A General Equilibrium Analysis of Technology, Migration and Deforestation in the Brazilian Amazon." In A. Angelsen and D. Kaimowitz, eds. *Agricultural Technologies and Tropical Deforestation.* CIFOR, CABI Publishers.

CEPEA/ESALQ/USP 2003. "Cadeia Agroindustrial de Carne Bovina." In *Indicadores CEPEA* (CEPEA website: http://cepea.esalq.usp.br).

Chomitz, K. and T.S. Thomas. 2000. "Geographic Patterns of Land Use and Land Intensity." World Bank, Development Research Group, Draft Paper, Washington, D.C.

Costa, F.G. 2000. "Avaliação do Potencial de Expansão da Soja na Amazônia Legal: uma Aplicação do Modelo de Von Thünen." Dissertação de Tese de Mestrado, Escola Superior de Agricultura Luiz de Queiroz (ESALQ), Universidade de São Paulo, Piracicaba.

Cropper, M., C. Griffith, and M. Muthukumara. 1997. "Roads, Population Pressures and Deforestation in Thailand, 1976-89." Policy Research Working Paper 1726, World Bank, Washington, D.C.

Diewald, C. 2002. "Notas sobre Políticas Públicas para o Brasil: Opções de Desenvolvimento e Conservação para as Florestas do Brasil." Banco Mundial, Brasília.

Faminow, M.D. 1998. *Cattle, Deforestation, and Development in the Amazon: an Economic, Agronomic and Environmental Perspective.* Oxford University Press.

Faminow, M.D. and C. Dahl. 1999. *Smallholders, Cattle and the Internal Drivers of Deforestation in the Western Brazilian Amazon.* Department of Agricultural Economics, University of Manitoba, Winnipeg, Canada.

Fearnside, P.M. 1993. "Deforestation in the Brazilian Amazonia: The Effect of Population and Land Tenure." Ambio 22(8):537-45.

Fearnside, P. 1997. "Environmental Services as a Strategy for Sustainable Development in Rural Amazônia." *Ecological Economics* 20:53-70.

Fearnside, P.M. 2001. "O cultivo da Soja como Ameaça para o Meio Ambiente na Amazônia Brasileira." In L. Forline and R. Murrieta, eds. *Amazônia 500 Anos: o V Centenário e o Novo Milênio: Lições de História e Reflexões para uma Nova Era.* Museu Paraense Emílio Goeldi, Belém.

Fearnside, P.M. 2002. "Controle de Desmatamento no Mato Grosso: um Novo Modelo para Reduzir a Velocidade da Perda de Floresta Amazônica." Paper presented in the Seminar "Aplicações do Sensoriamento Remoto e de Sistemas de Informação Geográfica no Monitoramento e Controle do Desmatamento na Amazônia Brasileira." Ministry of Environment, Brasília.

Ferraz, C. 2001. "Explaining Agriculture Expansion and Deforestation: Evidence from the Brazilian Amazon—1980/98." Texto Para Discussão No.282, IPEA, Rio de Janeiro.

Food and Agricultural Organization of the United Nations (FAO). 1981. "Los Recursos Forestales de la América Tropical." 32/6. Technical Report No. 1, Rome.

Fundação Getúlio Vargas. 1999. *Desenvolvimento e Implantação de Projetos Relacionados ao Programa de Ações Estratégicas da SUDAM – 1998/1999*. Rio de Janeiro.

Gasques, J.G. 2001. "Gastos Públicos na Agricultura." Texto para Discussão 782, IPEA, Brasília.

GEIPOT. 2002. Web Site www.transportes.gov.br/bit/trodo/estatistica.

Haan, Cornelis de, et al. 2001. *Livestock Development: Implications for Rural Poverty, the Environment, and Global Food Security*. Washington D.C.: World Bank.

Haddad, P. and F.A. Rezende. 2001. *Instrumentos Econômicos para o Desenvolvimento Sustentável da Amazônia*. Ministério do Meio Ambiente, Secretaria de Coordenação da Amazônia, Brasília.

Hecht, S. 1982. "Cattle Ranching Development in the Eastern Amazon: Evaluation of Development Strategy." Ph.D. Dissertation. Berkeley, University of California.

Hecht, S. 1993. "The Logic of Livestock and Deforestation in Amazonia." *Bioscience* 43:687-695.

Hecht, S.B., R. Norgaard, and G. Possio. 1988. "The Economics of Cattle Ranching in Eastern Amazonia." *Interciencia* 13:233-240.

Homma, A.K.O. 1993. "Expansão da Fronteira Agrícola na Amazônia: Lucros Decorrem da Especulação ou do Processo Produtivo?" In *Extrativismo Vegetal na Amazônia: limites e oportunidades*. Belém: EMBRAPA-Amazônia Oriental; Brasília: EMBRAPA-SPI.

Homma, A.K.O., R.T. Walker, F.N. Scatena, A.J. Conto, R.A. Carvalho, C.A.P. Ferreira, and A.I.M. Santos. 1995. "Reducão dos Desmatamentos na Amazõnia: Política Agrícola ou Ambiental?" Paper presented at the XXXIII Congresso Brasileiro de Economia e Sociologia Rural. Curitiba, Paraná.

Horton, B. et al. 2002. "Evaluating Non-users Willingness to Pay for the Implementation of a Proposed National Parks Program in Amazonia." A UK/Italian Contingent Valuation Study by CSERGE WP ECM 02-01.

Instituto Brasileiro de Geografia e Estatística (IBGE). Various years. Yearbooks, Agricultural Censuses.

Instituto Nacional de Pesquisas Espaciais (INPE). Various years. *Monitoramento da Floresta Amazônica Brasileira*. São Paulo.

Instituto Nacional de Pesquisas Espaciais (INPE). 2002. *Monitoring of the Brazilian Amazon Forest by Satellite 2000-2001*. Brasília, INPE, FUNCATE.

Kaimowitz, D. and A. Angelsen. 1998. *Economic Models of Tropical Deforestation—A Review*. Center for International Forestry Research (CIFOR), Indonesia.

Laurance, W.F., M.A. Cochrane, S. Bergen, P.M. Fearnside, P. Delamônica, C. Barber, S. D'Angelo, and T. Fernandes. 2001. "The Future of the Brazilian Amazon." Science 291 : 438-439.

Mahar, D. 1989. *Government Policies and Deforestation in Brazil's Amazon Region*. Washington, D.C.: World Bank.

Margulis, S. 2000. "Who are the Agents of Deforestation in the Amazon and Why do They Deforest?" Concept Paper, World Bank, Brasília, 2000.

Menezes, M.A. 2001. "O Controle Qualificado do Desmatamento e o Ordenamento Territorial na Região Amazônica." In *Causas e Dinâmica do Desmatamento na Amazônia*. Ministério do Meio Ambiente. Brasília.

Mertens, B., R. Poccard-Chapuis, M.-G. Piketty, A.-E. Lacques, and A. Venturieri. 2002. "Crossing Spacial Analyses and Livestock Economics to Understand Deforestation in the Brazilian Amazon: the Case of São Félix do Xingú in South Pará." *Agricultural Economics*, 27: 269-294.

Moreira, A. and E.J. Reis. 2002. "Determinantes e Tendências da Ocupação da Amazônia: um Modelo Econométrico." Background paper to the present study. Rio de Janeiro.

Mueller, C. 2002. Working Paper prepared for World Bank on Cerrado.

Mueller, C.C. 1977. "Pecuária de Corte no Brasil Central—Resultado das simulações com modelos de programação linear." In *Revista de Economia Rural*, SOBER. São Paulo, v.2.

Nepstad, D.C., A.G. Moreira, and A.A. Alencar. 1999. "Flames in the Rainforest: Origins, Impacts and Alternatives to Amazonian Fire." Pilot Program to Conserve the Brazilian Rainforest. Brasília.

Nepstad, D., J.P. Capobianco, A.C. Barros, G. Carvalho, G. Murtinho, U. Lopes, and P. Lefebvre. 2000. *Avança Brasil: os Custos Ambientais para a Amazônia*. Belém.

Pacheco, P. 2002a. "Deforestation in the Brazilian Amazon: A Review of Estimates at the Municipal Level." Background paper to the present study. Belém.

Pacheco, P. 2002b. "Revisiting the Role of Fiscal Incentives on Driving Livestock Expansion in the Brazilian Amazon." Draft for Discussion, IPAM/CGIAR, Belém.

Pearce, D. W. 1991. "An Economic Approach to Saving the Tropical Forests." In D. Helm, ed. *Economic Policy Towards the Environment*. Oxford: Blackwell.

Pearce, D.W. 1993. *Economic Values and the Natural World*. London: Earthscan Publications Limited.

Pfaff, A.S. 1997. "What Drives Deforestation in the Brazilian Amazon: Evidence from Satellite and Socioeconomic Data." *Journal of Environmental Economics and Management* 37:26-43.

Redwood, J. 2002. "World Bank Approaches to the Brazilian Amazon: the Bumpy Road Towards Sustainable Development." Sustainable Development Working Paper 13, Latin America and Caribbean Region, World Bank, Washington, D.C.

Reis, E.J. and S. Margulis. 1991. "Options for Slowing Amazon Jungle Clearing." In, Rudiger Dornbusch and James M. Poterba, eds. *Global Warming: Economic Policy Responses*. Cambridge, MIT Press.

Santana, A.C. 2000. *Agregação de Valor na Cadeia Produtiva da Pecuária de Corte no Estado do Pará*. FCAP, Belém.

Schneider, R. 1991. "An Analysis of Environmental Problems in the Amazon." Report No. 9104-BR, World Bank, Washington, D.C.

Schneider, R. 1995. "Government and the Economy on the Amazon Frontier." World Bank Environment Paper Number 11, Washington, D.C.

Schneider, R., E. Arima, A. Veríssimo, P. Barreto, and C. Souxa, Jr. 2000. *Amazônia Sustentável: Limitantes e Oportunidades para o Desenvolvimento Rural*. Séries Parcerias Banco Mundial – Brasil, and IMAZON.

Segura-Bonilla, O. 2000. "Forestry Policy in Costa Rica." In M. Dore, and R. Guevara, eds. *Sustainable Forest Management and Global Climate Change: Selected Case Studies form the Americas*. Edward Elgar Publishing, Cheltenham.

Seroa da Motta, R.S. 2002. "Estimativa do Custo Econômico do Desmatamento na Amazônia." Background paper to the present study. Rio de Janeiro.

Seroa da Motta, R., D. Nepstad, and M.J.C. Mendonça. 2001. "Custo Econômico do Uso do Fogo na Amazônia." Mimeo. Rio de Janeiro: IPEA/Ipam.

Seroa da Motta, R., J.M.D. Oliveira, and S. Margulis. 2000. "Proposta de Tributação Ambiental na Atual Reforma Tributária Brasileira." Texto para Discussão No.738, Rio de Janeiro.

Skole, D. and C. Tucker. 1993. "Tropical deforestation and Habitat Fragmentation in the Amazon: Satellite Data from 1978 to 1988." *Science* 260:1905-1910.

Superintendência do Desenvolvimento da Amazônia (SUDAM). 1995. *Avaliação da Política de Incentivos*